# Linear Cryptanalysis

Applications of cryptography are plenty in everyday life. This guidebook is about the security analysis or "cryptanalysis" of the basic building blocks on which these applications rely. Rather than covering a variety of techniques at an introductory level, this book provides a comprehensive and in-depth treatment of linear cryptanalysis. The subject is introduced from a mathematical point of view, providing an overview of the most influential papers on linear cryptanalysis and placing them in a consistent framework based on linear algebra. A large number of examples and exercises are included, drawing upon practice as well as theory.

The book is accessible to students with no prior knowledge of cryptography. It covers linear cryptanalysis starting from the basics, including linear approximations and trails, correlation matrices, automatic search and key-recovery techniques, to advanced topics such as multiple and multidimensional linear cryptanalysis, zero-correlation approximations, and the geometric approach.

TIM BEYNE is a postdoctoral researcher at Katholieke Universiteit (KU) Leuven, Belgium. He completed his dissertation in 2023 under the supervision of Vincent Rijmen. He is currently a junior postdoctoral fellow of the FWO, Belgium. His research has been recognized with awards at Asiacrypt 2018, Crypto 2021 and Asiacrypt 2021, and with the 2024 Nokia Bell Scientific Award.

VINCENT RIJMEN has been a professor at KU Leuven, Belgium, since 2007, and Adjunct Professor with the University of Bergen, Norway, since 2019. He is co-designer of the Advanced Encryption Standard (AES). He obtained his PhD on the design and cryptanalysis of block ciphers in 1997 and has been publishing articles at international cryptology conferences since 1993.

"Eleven chapters, written by two renowned international experts, brilliantly and smoothly present the fundamentals of linear cryptanalysis. This valuable book covers all the concepts and tools useful to students. Extremely well written, it also provides the necessary material for researchers wishing to deepen their knowledge of linear attacks and their extensions, and spreads the state of the art with novel features. Indispensable for all students and researchers interested in the subject, it will undoubtedly become an essential reference work."

— *Claude Carlet, University of Paris 8 (Emeritus) and University of Bergen*

"Vincent Rijmen and Tim Beyne have significantly advanced the field of linear cryptanalysis through their collaborative research. Their work has introduced innovative methodologies and deepened our understanding of cryptographic security."

— *Lars R. Knudsen, University of South Denmark*

"Beyne and Rijmen have done a terrific job in their book *Linear Cryptanalysis*, focusing on one of the most prominent – and effective – symmetric-key cryptanalytic methods. The basic idea behind linear cryptanalysis can be quite intuitive, yet its application in practice requires rigorous mathematical and statistical tools, which they cover in a precise manner in the monograph. The chapters on computing linear trails, deriving the corresponding correlations, and estimating the cost of attacks provide a rigorous foundation for understanding linear cryptanalysis; the discussion of key-recovery techniques is likely to be also useful when considering other cryptanalytic methods; and the different variants and extensions of linear cryptanalysis offer an illustration of how the ideas behind cryptanalytic methods can be broadened to apply to more ciphers. I particularly like the final chapters, which provide a very original and powerful algebraic representation of linear cryptanalysis, based on the recent work of one of the authors. The writing is clear, and the historic notes and exercises are great. This is a very nice book on cryptanalysis, which should be of interest to both novice and experienced cryptographers."

— *Carlos Cid, Simula UiB and Okinawa Institute of Technology*

"The initiative and effort of Professors Tim Beyne and Vincent Rijmen in writing this textbook on linear cryptanalysis is highly appreciated. This is timely and important. Cryptanalysis is a very important aspect of cryptology, and it all started with the introduction of linear cryptanalysis by Matsui. It has gained importance over time. The authors have made a comprehensive presentation on this subject. The book may be used as a text as well as a reference for advanced research. The main tool is correlation between two binary strings and its statistical analysis. The authors have handled all the issues thoroughly in such a way that it will be useful to both beginners and experts. In addition to a statistical approach, the authors have provided a geometric approach. I hope the book becomes popular with the global crypto community, especially students and young researchers."

— *Bimal Kumar Roy, Indian Statistical Institute*

# Linear Cryptanalysis

**TIM BEYNE**
*KU Leuven*

**VINCENT RIJMEN**
*KU Leuven and University of Bergen*

Shaftesbury Road, Cambridge CB2 8EA, United Kingdom

One Liberty Plaza, 20th Floor, New York, NY 10006, USA

477 Williamstown Road, Port Melbourne, VIC 3207, Australia

314–321, 3rd Floor, Plot 3, Splendor Forum, Jasola District Centre, New Delhi – 110025, India

103 Penang Road, #05–06/07, Visioncrest Commercial, Singapore 238467

Cambridge University Press is part of Cambridge University Press & Assessment, a department of the University of Cambridge.

We share the University's mission to contribute to society through the pursuit of education, learning and research at the highest international levels of excellence.

www.cambridge.org
Information on this title: www.cambridge.org/9781009607865
DOI: 10.1017/9781009607872

© Tim Beyne and Vincent Rijmen 2026

This publication is in copyright. Subject to statutory exception and to the provisions of relevant collective licensing agreements, no reproduction of any part may take place without the written permission of Cambridge University Press & Assessment.

When citing this work, please include a reference to the DOI 10.1017/9781009607872

First published 2026

*A catalogue record for this publication is available from the British Library*

Library of Congress Cataloging-in-Publication Data
Names: Beyne, Tim author | Rijmen, Vincent, 1970– author
Title: Linear cryptanalysis / Tim Beyne, Vincent Rijmen.
Description: Cambridge ; New York, NY : Cambridge University Press, 2026. | Includes bibliographical references and index.
Identifiers: LCCN 2025025968 (print) | LCCN 2025025969 (ebook) | ISBN 9781009607865 hardback | ISBN 9781009607889 paperback | ISBN 9781009607872 epub
Subjects: LCSH: Cryptography | Ciphers | Correlation (Statistics)
Classification: LCC QA268 .B478 2026 (print) | LCC QA268 (ebook)
LC record available at https://lccn.loc.gov/2025025968
LC ebook record available at https://lccn.loc.gov/2025025969

ISBN 978-1-009-60786-5 Hardback

Cambridge University Press & Assessment has no responsibility for the persistence or accuracy of URLs for external or third-party internet websites referred to in this publication and does not guarantee that any content on such websites is, or will remain, accurate or appropriate.

For EU product safety concerns, contact us at Calle de José Abascal, 56, 1°, 28003 Madrid, Spain, or email eugpsr@cambridge.org

# Contents

|      | *Preface*                                           | *page* ix |
|------|-----------------------------------------------------|-----------|
| **1** | **Introduction**                                   | 1         |
| 1.1  | Cryptographic primitives                            | 1         |
| 1.2  | Linear approximations                               | 4         |
| 1.3  | Linear trails and the piling-up lemma               | 7         |
| 1.4  | Recovering a key                                    | 9         |
| 1.5  | Remaining problems                                  | 13        |
| 1.6  | Historical remarks                                  | 13        |
| 1.7  | References                                          | 14        |
| 1.8  | Exercises                                           | 14        |
| **2** | **Correlation matrices**                           | 17        |
| 2.1  | Correlation of a random variable on $\mathbb{F}_2$  | 17        |
| 2.2  | Correlation between Boolean functions               | 18        |
| 2.3  | Correlation matrices                                | 19        |
| 2.4  | Correlation matrices of structured functions        | 21        |
| 2.5  | Linear trails                                       | 24        |
| 2.6  | Historical remarks                                  | 26        |
| 2.7  | References                                          | 28        |
| 2.8  | Exercises                                           | 28        |
| **3** | **Optimization of linear trails**                  | 34        |
| 3.1  | Branch and bound                                    | 34        |
| 3.2  | Mixed-integer linear programming                    | 38        |
| 3.3  | Satisfiability and satisfiability modulo theories   | 44        |
| 3.4  | Historical remarks                                  | 47        |
| 3.5  | References                                          | 48        |
| 3.6  | Exercises                                           | 48        |

| | | |
|---|---|---|
| **4** | **Statistics of linear cryptanalysis** | **52** |
| 4.1 | Statistical inference | 52 |
| 4.2 | Key-recovery using statistical hypothesis testing | 56 |
| 4.3 | Sampling strategies | 61 |
| 4.4 | Key-recovery using key ranking | 61 |
| 4.5 | Historical remarks | 63 |
| 4.6 | References | 63 |
| 4.7 | Exercises | 63 |
| **5** | **Key-recovery techniques** | **64** |
| 5.1 | Key-recovery using Algorithm 2 | 64 |
| 5.2 | Matsui's approach | 65 |
| 5.3 | Fast Fourier transformation method | 68 |
| 5.4 | Historical remarks | 72 |
| 5.5 | References | 73 |
| 5.6 | Exercises | 73 |
| **6** | **Multiple linear cryptanalysis** | **74** |
| 6.1 | Multiple linear cryptanalysis | 74 |
| 6.2 | Multidimensional linear cryptanalysis | 81 |
| 6.3 | Closing remarks | 85 |
| 6.4 | Historical remarks | 86 |
| 6.5 | References | 87 |
| 6.6 | Exercises | 88 |
| **7** | **Optimal statistical testing** | **91** |
| 7.1 | Probability measures | 91 |
| 7.2 | Simple hypotheses | 92 |
| 7.3 | Composite hypotheses | 98 |
| 7.4 | Optimal key-recovery | 104 |
| 7.5 | Historical remarks | 105 |
| 7.6 | References | 105 |
| 7.7 | Exercises | 106 |
| **8** | **Zero-correlation approximations** | **107** |
| 8.1 | The idea | 107 |
| 8.2 | Finding approximations with correlation zero | 108 |
| 8.3 | Using zero-correlation approximations | 111 |
| 8.4 | Statistical approach | 115 |
| 8.5 | Historical remarks | 116 |
| 8.6 | References | 116 |
| 8.7 | Exercises | 116 |

| | | |
|---|---|---|
| **9** | **Miscellaneous extensions** | 120 |
| 9.1 | Exact properties | 120 |
| 9.2 | Approximate properties | 126 |
| 9.3 | Historical remarks | 128 |
| 9.4 | References | 129 |
| 9.5 | Exercises | 130 |
| **10** | **Functions on Abelian groups** | 133 |
| 10.1 | Linear algebra over $\mathbb{C}$ | 133 |
| 10.2 | Fourier analysis on finite Abelian groups | 140 |
| 10.3 | Historical remarks | 146 |
| 10.4 | References | 146 |
| 10.5 | Exercises | 147 |
| **11** | **Geometric approach** | 150 |
| 11.1 | Geometric viewpoint | 150 |
| 11.2 | Linear cryptanalysis | 155 |
| 11.3 | Exact propagation | 158 |
| 11.4 | Approximate propagation | 160 |
| 11.5 | Historical remarks | 163 |
| 11.6 | References | 163 |
| 11.7 | Exercises | 164 |
| **Appendix A** | **Normal distribution** | 168 |
| A.1 | Univariate normal distribution | 168 |
| A.2 | Multivariate normal distribution | 169 |
| **Appendix B** | **Statistical formulary** | 171 |
| **Appendix C** | **List of block ciphers** | 173 |
| | *References* | 174 |
| | *Index* | 178 |

# Preface

Cryptanalysis remains a young and fast-evolving field. One consequence is that lecturers and thesis advisors often struggle to find adequate textbooks to teach students about the theory and practice of the field in an efficient way. With this book, we hope to fill this gap for linear cryptanalysis.

In our view, among all methods of symmetric-key cryptanalysis, linear cryptanalysis is the most suitable for a first course. It is not too difficult to acquire some intuition about how linear attacks are constructed and executed. At the same time, some basic and a few advanced concepts from linear algebra can be used to describe linear cryptanalysis in a scientific way.

## Scope of this book

If you study this book with care, you can expect to obtain a solid understanding of the basic theory of linear cryptanalysis, and familiarity with its most important extensions (multiple and multidimensional linear cryptanalysis, zero-correlation linear cryptanalysis, ...). If you also diligently complete the exercises, you will be able to apply this knowledge in practice. In due time, you will have no trouble understanding the current literature.

Nevertheless, in a book of this length, it is impossible to provide a complete overview of all related work, or even to reference all of it. The focus is decidedly on core cryptanalytic results, with many important but tangential topics (such as connections with linear codes, Boolean functions, ...) deferred to the exercises. Most of the examples and exercises are related to block ciphers, but we expect that you will be able to apply linear cryptanalysis to other cryptographic primitives, such as stream ciphers, when you encounter them.

In an attempt to avoid being accused of unwarranted favoritism, we have deliberately worked with short lists of reference works. These can be found at the end of each chapter. The selection of references is motivated from a historical point of view and does not necessarily reflect the best place to learn more about the material.

Apart from linear cryptanalysis, there are many other important cryptanalytic techniques. A few are mentioned in this book, but only when we saw a way to connect them to our treatment of linear cryptanalysis. Several important cryptanalytic techniques are therefore not discussed.

## Using this book as a beginner

This book is based on a one-semester course on linear cryptanalysis, first taught at KU Leuven in the fall of 2023. This was a first course in cryptanalysis, aimed at master's students with a background in mathematics and mathematical engineering. This textbook can serve as the basis for similar courses, or it can be read linearly for self-studying. However, the book can also be used piecemeal, and we give some suggestions for this below.

Chapters 1 to 5 of this book can be used to cover the basic principles of linear cryptanalysis, e.g., as part of a broader course on cryptanalysis. Chapters 6 to 9 cover more advanced topics, suitable if one wants to expand both in depth and in breadth, up to the state of the art. Chapters 10 and 11 discuss the current state of the art; we recommend them to those who intend to study other cryptanalytic techniques, such as differential and integral cryptanalysis, as well as to aspiring researchers.

For short courses of one or two lectures, we recommend against relying exclusively on Chapter 1. This chapter raises more questions than it answers. For the same reason, beginners should quickly move on to Chapter 2, rather than to try and understand every last detail in Chapter 1 on first reading.

Readers who are mainly interested in the mathematical aspects of linear cryptanalysis, more than the cryptanalysis of actual ciphers, should not dwell too long on Chapter 1 and can safely skip Chapters 3, 5 and 9.

## Using this book as an expert

As researchers and reviewers we sometimes encounter work using outdated methods or overly simplistic approximations. With this book, we hope to help spread the state of the art. Experts may find this book useful as a reference, or because of some of its novel features that are outlined below.

Early on, in Chapter 2, we emphasize the description of linear cryptanalysis using correlation matrices. In our view, this is the most efficient way to obtain the main results without raising the level of abstraction too much. Sooner or later, correlation matrices become indispensable in any case, and introducing them earlier makes for a smoother transition to Chapters 10 and 11.

The discussion of statistical aspects of linear cryptanalysis is a delicate matter. On the one hand, explicit formulas are helpful to understand the main factors that influence the cost of an attack. On the other hand, in the interest of accuracy, it is desirable to use as few simplifications as possible. We have tried to strike a balance by giving closed-form formulas where possible without getting bogged down in technical details, emphasizing in each case the simplifications that are made. One should keep in mind that the most significant approximations in Chapters 4 and 7 are related to statistical modeling of reality (sampling strategy, key-dependence of correlations, wrong key randomization, ...), rather than to mathematical issues such as convergence rates of limit theorems.

Our exposition of multidimensional linear cryptanalysis in Chapter 6 is somewhat original, in the sense that it does not depend on a choice of basis for the space of masks. This approach to multidimensional linear approximations is taken further in Chapter 11.

Chapter 11 introduces the geometric approach to cryptanalysis from a relatively concrete point of view, with an emphasis on linear cryptanalysis and some closely related techniques. A more general treatment would have required some mathematical background beyond linear algebra. Nevertheless, we tried to present the results and examples in a way that is consistent with the general theory, of which a few exercises provide a glimpse. For example, from the start, we insist on the difference between $\mathbb{C}[G]$ and $\mathbb{C}^G$ – but we do not discuss the coalgebra and algebra structure of these spaces.

# 1
# Introduction

Applications of cryptography are plenty and can be found in everyday life. This book is not about these applications, but about the basic building blocks on which their security is based. These building blocks are called *cryptographic primitives*, and this book introduces you to their security analysis. Rather than covering a variety of techniques at an introductory level, we will study a single family of techniques – linear cryptanalysis – in depth.

This chapter introduces linear cryptanalysis from the point of view that historically led to its discovery. This "original" description has the advantage of being concrete, but it is not very effective. However, it raises important questions that motivate later chapters.

## 1.1 Cryptographic primitives

Given the digital nature of modern cryptography, most primitives operate on fixed-length strings of bits. Throughout this book, the set of bitvectors of length $n$ is denoted by $\mathbb{F}_2^n$ – where $\mathbb{F}_2$ is the field of integers modulo two. *Block ciphers* are the most well-known primitives. A block cipher with block size $n$ is a family of invertible functions from $\mathbb{F}_2^n$ to $\mathbb{F}_2^n$. A function from the family is denoted by $\mathsf{E}_k$, where the index $k$ – usually a bitvector – is called the *key*.

### 1.1.1 Analysis

In most applications of block ciphers, the key is kept secret. The security of a block cipher is then related to how difficult it is for an adversary to determine (*recover*) its key. However, the definition of block cipher security can be – and often is – generalized. For example, given the ability to query the block cipher, one can consider how difficult it is to determine (*distinguish*) whether one is interacting with the cipher or with a dummy algorithm that returns random results.[1]

The *difficulty* of an attack includes several aspects, beginning with the properties of the algorithm that implements the attack: its running time, memory requirements, degree of parallelism, success probability, ... One should also take into account the amount and type of information that is required. In a known-plaintext attack, some input-output pairs for the primitive are available – the inputs are sampled from a known distribution. In a chosen-plaintext attack, the inputs are chosen by the attacker.

There is a simple attack strategy that works for every block cipher: *exhaustive key search*. It requires a few known plaintext-ciphertext pairs $(x_1, y_1), \ldots, (x_q, y_q)$ and works by looping through all possible values of the key, checking for every key $k$ whether or not $y_i = \mathsf{E}_k(x_i)$ for $i = 1, \ldots, q$. Exhaustive key search requires little memory and can easily be parallelized. It is often used as a benchmark to assess the relevance of other attacks: in order to qualify as an attack, an algorithm should outperform exhaustive search in at least one aspect.

### 1.1.2 Design

Ciphers can be constructed by composing relatively simple functions:

$$\mathsf{E}_k = \mathsf{R}_k^{(r)} \circ \cdots \circ \mathsf{R}_k^{(1)}.$$

"Relatively simple" usually means that the functions $\mathsf{R}_k^{(1)}, \ldots, \mathsf{R}_k^{(r)}$ can be efficiently evaluated on the target platform(s), and that they have a compact mathematical description that is well understood. All modern block ciphers follow this approach.

Iterated ciphers are ciphers where the functions $\mathsf{R}_k^{(i)}$ are instances of a single keyed family of functions:

$$\mathsf{E}_k = \mathsf{R}_{k_r} \circ \cdots \circ \mathsf{R}_{k_1}.$$

The functions $\mathsf{R}_{k_i}$ are called the *rounds* of $\mathsf{F}_k$. The sequence $(k_1, \ldots, k_r)$ is called the *expanded key* of the block cipher. It is derived by applying a function called the *key schedule* to the key $k$.

Key-alternating ciphers are iterated ciphers where the round function is the composition of a key-independent function and a key addition (in $\mathbb{F}_2^n$):

$$\mathsf{R}_{k_i}(x) = \mathsf{R}(x) + k_i.$$

---

[1] The results should be consistent with the fact that the cipher is a permutation.

## 1.1 Cryptographic primitives

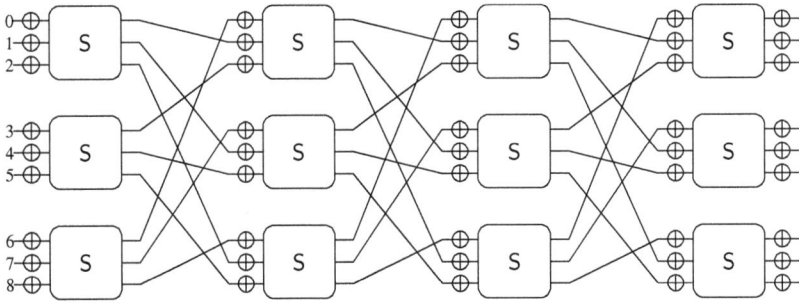

Figure 1.1 A block cipher with a block size of nine bits and four rounds.

The terms "iterated cipher" and "key-alternating cipher" are often used flexibly: even if the first or last round deviate slightly from the others, a cipher can still be called iterated or key-alternating.

Figure 1.1 depicts a block cipher with block size $n = 9$. It is used as a running example throughout this chapter. The cipher is a substitution-permutation network with a 45-bit key $k$ in $\mathbb{F}_2^{45}$. The round function consists of the following three operations:

**S-box layer.** This operation applies $S: \mathbb{F}_2^3 \to \mathbb{F}_2^3$ to three groups of state bits:

$$(x_8, \ldots, x_0) \mapsto S(x_8, x_7, x_6) \| S(x_5, x_4, x_3) \| S(x_2, x_1, x_0),$$

where the symbol "$\|$" denotes concatenation of bitvectors. The S-box function S was first used in the block cipher 3-Way and is defined by the following lookup table. Appendix C contains a list of all ciphers mentioned in this book, including 3-Way.

| $x$ | 000 | 001 | 010 | 011 | 100 | 101 | 110 | 111 |
|---|---|---|---|---|---|---|---|---|
| $S(x)$ | 111 | 010 | 100 | 101 | 001 | 110 | 011 | 000 |

**Bit permutation.** The second operation shuffles the bits of the state by mapping the $i$th output bit of S-box $j$ to input bit $j + 1$ (mod 3) of S-box $i$. Concretely, $(x_8, \ldots, x_0) \mapsto (x_5, x_2, x_8, x_4, x_1, x_7, x_3, x_0, x_6)$.

**Key addition.** Each round ends by adding a round key to the state. In the $i$th round (counting from one), the key addition operation corresponds to the function $(x_8, \ldots, x_0) \mapsto (x_8 + k_{9i+8}, \ldots, x_0 + k_{9i})$. In Figure 1.1, the key addition is represented by the "$\oplus$" symbol.

After adding the key bits $(k_8, \ldots, k_0)$ to the plaintext, the cipher successively evaluates the above operations four times.

To avoid misunderstandings, it should be emphasized that this example is a textbook cipher and should not be used in practice. Its small key size of 45 bits makes exhaustive search feasible (as there are only $2^{45}$ possible keys), and even more efficient attacks will be given later in this chapter. It is also worth mentioning that most real-world ciphers have much larger block sizes. For example, the *Advanced Encryption Standard* (AES) has a block size of 128 bits.

## 1.2 Linear approximations

Linear cryptanalysis is based on *linear approximations*. These are probabilistic linear relations between the input and output bits of a function $\mathsf{F}: \mathbb{F}_2^n \to \mathbb{F}_2^m$. By probabilistic, we mean that the relation does not hold for all input values of the function. By linear, we mean linear over the field $\mathbb{F}_2$. If $y = \mathsf{F}(x)$, then a linear approximation corresponds to an equation of the form

$$\sum_{i=1}^m v_i \, y_i = \sum_{i=1}^n u_i x_i \, .$$

This can be written more compactly as $v^\mathsf{T} \mathsf{F}(x) = u^\mathsf{T} x$, with $u$ and $v$ vectors with coordinates $(u_1, \ldots, u_n)$ and $(v_1, \ldots, v_m)$, respectively. Sometimes, we regard $u$ and $v$ as bitstrings. The vectors $u$ and $v$ are called input and output masks, respectively. Since the masks $u$ and $v$ determine the approximation, we say that a linear approximation *is* a pair of masks $(u, v)$ in $\mathbb{F}_2^n \times \mathbb{F}_2^m$.

### 1.2.1 Bias

Let **x** be a uniformly distributed random variable taking values in $\mathbb{F}_2^n$. Consider the probability of a linear approximation $(u, v)$ of $\mathsf{F}$:

$$\Pr_{\mathbf{x}}\left[u^\mathsf{T} \mathbf{x} = v^\mathsf{T} \mathsf{F}(\mathbf{x})\right] = \frac{\left|\{x \in \mathbb{F}_2^n \mid u^\mathsf{T} x = v^\mathsf{T} \mathsf{F}(x)\}\right|}{2^n} \, .$$

If the above probability is $1/2$, then $u^\mathsf{T} x$ and $v^\mathsf{T} \mathsf{F}(x)$ are unrelated: for half of the inputs $x$ they have the same value, and for the other half they have complementary values. Based on this observation, the *bias* $\epsilon_{u,v}$ of a linear approximation $(u, v)$ of $\mathsf{F}$ is defined as

$$\epsilon_{u,v} = \Pr_{\mathbf{x}}\left[u^\mathsf{T} \mathbf{x} = v^\mathsf{T} \mathsf{F}(\mathbf{x})\right] - \frac{1}{2} \, .$$

If $\epsilon_{u,v} \neq 0$, then the linear approximation $(u, v)$ is called *effective*.

## 1.2 Linear approximations

*Example 1.1* Let S be the S-box from the example cipher defined in Section 1.1. Consider the approximation $(u, v) = (001, 011)$ of the function S. To compute the bias, construct the following table:

| $x$ | $u^T x$ | $S(x)$ | $v^T S(x)$ |
|---|---|---|---|
| 000 | 0 | 111 | 0 |
| 001 | 1 | 010 | 1 |
| 010 | 0 | 100 | 0 |
| 011 | 1 | 101 | 1 |
| 100 | 0 | 001 | 1 |
| 101 | 1 | 110 | 1 |
| 110 | 0 | 011 | 0 |
| 111 | 1 | 000 | 0 |

It follows that the bias of $(001, 011)$ equals $6/8 - 1/2 = 1/4$. Similarly, you can show as an exercise that the bias of the approximation $(100, 100)$ is $-1/4$. ▷

### 1.2.2 Linear approximation tables

The *linear approximation table* (LAT) of a function $F: \mathbb{F}_2^n \to \mathbb{F}_2^m$ is a table containing the biases of all linear approximations of F, scaled by a factor $2^n$. That is,

$$\mathsf{LAT}_{u,v} = 2^n \, \epsilon_{u,v} .$$

Including the null masks, there are $2^{n+m}$ approximations, and the LAT is a table with $2^n$ rows and $2^m$ columns. Observe that the entries of the LAT are indexed by bitvectors, i.e., elements of $\mathbb{F}_2^n$ and $\mathbb{F}_2^m$.

**Theorem 1.1** *Let F be a function from $\mathbb{F}_2^n$ to $\mathbb{F}_2^m$. The LAT of F has the following properties:*

1. $\mathsf{LAT}_{0,0} = 2^{n-1}$.
2. *For all nonzero $u$ in $\mathbb{F}_2^n$, $\mathsf{LAT}_{u,0} = 0$.*

*If F is invertible, then the LAT of F has the following additional properties:*

3. *For all nonzero $v$ in $\mathbb{F}_2^m$, $\mathsf{LAT}_{0,v} = 0$.*
4. *All entries of the LAT are even numbers.*

*Proof* The first property follows from $\epsilon_{0,0} = \Pr_{\mathbf{x}}[0 = 0] - 1/2 = 1/2$. For the second property, note that

$$\epsilon_{u,0} = \frac{|\{x \in \mathbb{F}_2^n \mid u^\mathsf{T} x = 0\}|}{2^n} - \frac{1}{2}.$$

For all $u \neq 0$, there are $2^{n-1}$ values $x$ in $\mathbb{F}_2^n$ such that $u^\mathsf{T} x = 0$. Hence, the first term above equals $1/2$ and the result is zero. If F is invertible, then $m = n$ and the third property follows by a similar argument. Indeed,

$$\epsilon_{0,v} = \frac{|\{x \in \mathbb{F}_2^n \mid v^\mathsf{T} F(x) = 0\}|}{2^n} - \frac{1}{2} = \frac{|\{y \in \mathbb{F}_2^n \mid v^\mathsf{T} y = 0\}|}{2^n} - \frac{1}{2},$$

where the second equality is due to the fact that F is invertible.

If $u = 0$ or $v = 0$, then the fourth property follows from properties (1) to (3). Otherwise, the functions $x \mapsto u^\mathsf{T} x$ and $x \mapsto v^\mathsf{T} F(x)$ both take the value 0 for exactly $2^{n-1}$ inputs. Let $a$ denote the number of values of $x$ such that $u^\mathsf{T} x = 0$ and $v^\mathsf{T} F(x) = 0$. This leads to the following partition of $\mathbb{F}_2^n$:

|  | $u^\mathsf{T} x = 0$ | $u^\mathsf{T} x = 1$ |
|---|---|---|
| $v^\mathsf{T} F(x) = 0$ | $a$ | $2^{n-1} - a$ |
| $v^\mathsf{T} F(x) = 1$ | $2^{n-1} - a$ | $a$ |

In particular, there are $2^{n-1} - a$ values of $x$ such that $u^\mathsf{T} x = 1$ and $v^\mathsf{T} F(x) = 0$. Since F is invertible, there are also $2^{n-1} - a$ values of $x$ such that $u^\mathsf{T} x = 0$ and $v^\mathsf{T} F(x) = 1$. It follows that there are $2^{n-1} - (2^{n-1} - a) = a$ values of $x$ such that $u^\mathsf{T} x = 1$ and $v^\mathsf{T} F(x) = 1$.

In conclusion, the number of $x$ such that $u^\mathsf{T} x = v^\mathsf{T} F(x)$ is equal to $2a$. □

*Example 1.2* The linear approximation table of S equals

$$\mathsf{LAT} = \begin{bmatrix} 4 & 0 & 0 & 0 & 0 & 0 & 0 & 0 \\ 0 & -2 & 0 & 2 & 0 & -2 & 0 & -2 \\ 0 & 0 & -2 & -2 & 0 & 0 & 2 & -2 \\ 0 & -2 & 2 & 0 & 0 & 2 & 2 & 0 \\ 0 & 0 & 0 & 0 & -2 & 2 & -2 & -2 \\ 0 & 2 & 0 & 2 & -2 & 0 & 2 & 0 \\ 0 & 0 & -2 & 2 & 2 & 2 & 0 & 0 \\ 0 & -2 & -2 & 0 & -2 & 0 & 0 & 2 \end{bmatrix}.$$

As an exercise, verify the properties listed in Theorem 1.1. ▷

## 1.3 Linear trails and the piling-up lemma

For some functions, the LAT can easily be determined analytically. Two examples are given below – both are used in the example cipher from Section 1.1.

**Addition with a constant.** Let F be the function that adds a constant $c$ to its input, i.e., $F(x) = x + c$. For all linear approximations $(u, v)$ of F, one has

$$\Pr_x\left[u^T x = v^T F(x)\right] = \Pr_x\left[u^T x = v^T x + v^T c\right] = \Pr_x\left[(u+v)^T x = v^T c\right].$$

If $u \neq v$, then the probability is $1/2$, and hence the bias is zero. If $u = v$, then the probability is one if $v^T c = 0$ and zero if $v^T c = 1$. Adopting the convention that $(-1)^b = 1$ for $b = 0$ in $\mathbb{F}_2$ and $(-1)^b = -1$ for $b = 1$ in $\mathbb{F}_2$, the bias is

$$\epsilon_{u,v} = \begin{cases} (-1)^{v^T c} \frac{1}{2} & \text{if } u = v, \\ 0 & \text{else.} \end{cases}$$

Although the key of a block cipher is secret, it is just a constant! Hence, the bias of a linear approximation over a key addition is given by the above formula.

**Bit permutation.** It is not difficult to see that if F is a bit permutation, i.e., it shuffles the bits of the input, then the probability of a linear approximation $(u, v)$ of F is one if $v = F(u)$ and $1/2$ otherwise. That is,

$$\epsilon_{u,v} = \begin{cases} \frac{1}{2} & \text{if } v = F(u), \\ 0 & \text{else.} \end{cases}$$

## 1.3 Linear trails and the piling-up lemma

This section turns to the problem of finding the bias of a linear approximation of a composition of functions $F = F_r \circ \cdots \circ F_1$, given only the biases of linear approximations of the individual functions $F_1, \ldots, F_r$. This question is particularly relevant when analyzing iterated ciphers.

Let $z_1$ be uniform random and $z_{i+1} = F_i(z_i)$ for $i = 1, \ldots, r$. To determine the bias $\epsilon_{u_1, u_{r+1}}$ of a linear approximation $(u_1, u_{r+1})$ of F, we consider successive linear approximations over the functions $F_1, \ldots, F_r$ such that the output mask of every approximation equals the input mask of the next approximation. The sequence of masks $(u_1, \ldots, u_{r+1})$ is called a *linear trail*. To determine $\epsilon_{u_1, u_{r+1}}$, define random variables $x_1, \ldots, x_r$ as follows:

$$\mathbf{x}_1 = u_1^T \mathbf{z}_1 + u_2^T \mathbf{z}_2$$
$$\mathbf{x}_2 = u_2^T \mathbf{z}_2 + u_3^T \mathbf{z}_3$$
$$\vdots$$
$$\underline{\mathbf{x}_r = u_r^T \mathbf{z}_r + u_{r+1}^T \mathbf{z}_{r+1}}$$
$$\sum_{i=1}^r \mathbf{x}_r = u_1^T \mathbf{z}_1 + u_{r+1}^T \mathbf{z}_{r+1} \,.$$

The bias of $\mathbf{x}_i$, i.e., $\Pr[\mathbf{x}_i = 0] - 1/2$, is equal to the bias of the linear approximation $(u_i, u_{i+1})$ of $F_i$. In general, the bias of $\sum_{i=1}^r \mathbf{x}_i$ cannot be determined from the biases of $\mathbf{x}_1, \ldots, \mathbf{x}_r$ alone. However, if $\mathbf{x}_1, \ldots, \mathbf{x}_r$ are independent, then it can be computed using the piling-up lemma.

**Lemma 1.2** (Piling-up lemma)  *Let $\mathbf{x}_1, \ldots, \mathbf{x}_r$ be random variables on $\mathbb{F}_2$ with biases $\epsilon_1, \ldots, \epsilon_r$. If $\mathbf{x}_1, \ldots, \mathbf{x}_r$ are independent, then the bias $\epsilon$ of the sum $\mathbf{x}_1 + \cdots + \mathbf{x}_r$ is*

$$\epsilon = 2^{r-1} \prod_{i=1}^r \epsilon_i \,.$$

*Proof* Consider the case $r = 2$. The bias of $\mathbf{x}_1 + \mathbf{x}_2$ satisfies

$$\frac{1}{2} + \epsilon = \left(\frac{1}{2} + \epsilon_1\right)\left(\frac{1}{2} + \epsilon_2\right) + \left(1 - \frac{1}{2} - \epsilon_1\right)\left(1 - \frac{1}{2} - \epsilon_2\right).$$

Working out the right-hand side above yields $\epsilon = 2\epsilon_1\epsilon_2$. The result for general $r$ follows by recursively applying the $r = 2$ case. $\square$

In the case of a linear trail, the random variables $\mathbf{x}_1, \ldots, \mathbf{x}_r$ are clearly not independent. Nevertheless, the piling-up lemma is used as a heuristic to estimate the bias of a linear approximation over a composition of functions:

$$\epsilon_{u_1, u_{r+1}} \approx 2^{r-1} \prod_{i=1}^r \epsilon_{u_i, u_{i+1}} \,.$$

Discussing the accuracy of this heuristic would lead us too far at this point. Instead, we defer this discussion to Chapter 2, where the formalism of correlation matrices will allow us to settle the issue in a simple way.

*Example 1.3* (Linear trail for the example cipher)  To find a nontrivial effective approximation over three rounds of the example cipher from Section 1.1, one can combine three effective one-round approximations.

Denote by **a** the (uniform random) input of the cipher, and by **b**, **c** and **d** the inputs of the first, second and third rounds. Finally, let **e** be the output of the third round. Using the LAT of S (in Example 1.2) and our earlier observations on additions with a constant and bit permutations, one can verify that

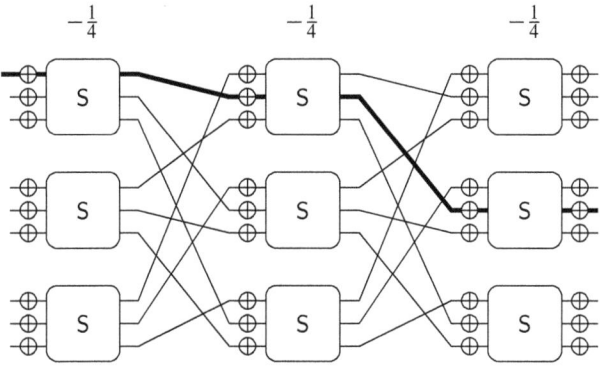

Figure 1.2 The linear trail from Example 1.3.

$\mathbf{a}_0 + \mathbf{b}_0 = 0$ with bias $\epsilon_0 = \frac{1}{2}(-1)^{k_0}$,
$\mathbf{b}_0 + \mathbf{c}_1 = 0$ with bias $\epsilon_1 = -\frac{1}{4}(-1)^{k_{10}}$,
$\mathbf{c}_1 + \mathbf{d}_4 = 0$ with bias $\epsilon_2 = -\frac{1}{4}(-1)^{k_{22}}$,
$\mathbf{d}_4 + \mathbf{e}_4 = 0$ with bias $\epsilon_3 = -\frac{1}{4}(-1)^{k_{31}}$.

The masks corresponding to the above trail are illustrated in Figure 1.2. In this figure, thick lines indicate nonzero bits of the masks. For example, the mask at the input of the third round is 000010000.

Heuristically, by the piling-up lemma and the identity $(-1)^x(-1)^y = (-1)^{x+y}$, the linear approximation $(000000001, 000010000)$ – or equivalently $\mathbf{a}_0 + \mathbf{e}_4 = 0$ – has bias

$$\epsilon \approx (-1)^{k_0+k_{10}+k_{22}+k_{31}+1}\frac{1}{16}.$$

As an exercise, try to find at least one other trail with the same input and output masks and show that the absolute value of its bias is smaller than 1/16. Since different trails yield different results, Lemma 1.2 should only be used for the trail that results in the highest bias. However, as shown in Chapter 2, even in this case there is no guarantee that the results are accurate. ▷

## 1.4 Recovering a key

An effective linear approximation of a block cipher can be used to set up a key-recovery attack. There are two basic methods to achieve this, called "Matsui's Algorithm 1" and "Matsui's Algorithm 2" after their originator.

### 1.4.1 Matsui's Algorithm 1

Matsui's Algorithm 1 recovers one bit of information about the expanded key of a key-alternating cipher.[2]

Consider a key-alternating block cipher $E_k = R_{k_r} \circ \cdots \circ R_{k_1}$ with $R_{k_i}(x) = R(x) + k_i$, and let $\epsilon_{u_i, u_{i+1}}$ denote the bias of the linear approximation $(u_i, u_{i+1})$ of R. The bias of the linear approximation $(u_i, u_{i+1})$ of $R_{k_i}$ then equals

$$(-1)^{u_{i+1}^T k_i} \epsilon_{u_i, u_{i+1}}.$$

In Section 1.3, the bias of a linear approximation was estimated using a linear trail. Let $(u_1, \ldots, u_{r+1})$ be a linear trail for the composition $E_k = R_{k_r} \circ \cdots \circ R_{k_1}$. Applying the piling-up lemma to this trail leads to the following estimate of the bias of the approximation $(u_1, u_{r+1})$:

$$\epsilon_{u_1, u_{r+1}} \approx 2^{r-1} \prod_{i=1}^{r} (-1)^{u_{i+1}^T k_i} \epsilon_{u_i, u_{i+1}} = 2^{r-1} (-1)^z \prod_{i=1}^{r} \epsilon_{u_i, u_{i+1}}.$$

On the right, $z$ is equal to $\sum_{i=1}^{r} u_{i+1}^T k_i$. This will be the bit of information that Matsui's Algorithm 1 recovers about the secret key. Observe that this is not a bit of the key in the strict sense of the word; it is a linear expression in some bits of the expanded key.

Provided that the approximative nature of the equation does not switch the sign, $z$ can be computed from the sign of $\prod_{i=1}^{r} \epsilon_{u_i, u_{i+1}}$ and $\epsilon_{u_1, u_{r+1}}$. The former can be determined from the theoretical analysis of the trail. The most likely value of the latter is obtained from the *empirical bias* of the linear approximation $(u_1, u_{r+1})$. The empirical bias is estimated using a random sample of plaintext-ciphertext pairs.

Given a random sample of $q$ plaintext-ciphertext pairs $(\mathbf{x}_i, \mathbf{y}_i)$, the empirical bias of the linear approximation $(u_1, u_{r+1})$ is

$$\hat{\epsilon} = \frac{1}{q} \left| \left\{ 1 \leq i \leq q \mid u_1^T \mathbf{x}_i = u_{r+1}^T \mathbf{y}_i \right\} \right| - \frac{1}{2}.$$

The average of $\hat{\epsilon}$ is $\epsilon_{u_1, u_{r+1}}$. For independent samples, the variance of the number of times the approximation holds is close to $q/4$. Indeed, for one sample, the variance is $(1/2 + \epsilon)(1 - 1/2 - \epsilon) \approx 1/4$. Hence, the standard

---

[2] Algorithm 1 is more widely applicable, but we only discuss the key-alternating case here.

## 1.4 Recovering a key

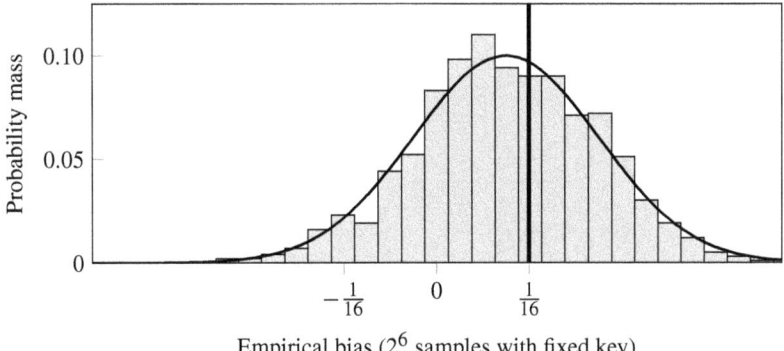

Empirical bias ($2^6$ samples with fixed key)

Figure 1.3 Histogram of the empirical bias for 1000 experiments.

deviation of $\hat{\epsilon}$ is approximately $1/\sqrt{4q}$. It follows that determining the sign of $\epsilon_{u_1,u_{r+1}}$ with high confidence requires a number of samples $q$ such that

$$\frac{1}{\sqrt{4q}} \ll |\epsilon_{u_1,u_{r+1}}|.$$

That is, Matsui's Algorithm 1 needs $q \gg 1/(2\epsilon_{u_1,u_{r+1}})^2$ samples.

*Example* 1.4 (Matsui's Algorithm 1 for the example cipher) Consider the linear approximation from Example 1.3. Using a linear trail, we estimated its bias as $-1/16 \cdot (-1)^z$ with $z = k_0 + k_{10} + k_{22} + k_{31}$. Based on the above, 64 samples should be enough to determine $z$. To test how well the attack works, we run it 1000 times (with key 00000001 01000000 00000000 00000000) and record the empirical bias. A histogram of the results is shown in Figure 1.3.

The average empirical bias over the 1000 experiments was slightly smaller than $1/16$. This is not a coincidence – the results in Chapter 2 imply that the bias is actually $3/64$ for the key used in the experiment. Simply taking the sign of the empirical correlation is most likely to yield $z = 0$ (the correct value). ▷

Linear cryptanalysis is usually called a known-plaintext attack, but observe that full plaintexts and ciphertexts are not needed to apply Matsui's Algorithm 1: The values of $u_1^T x_i$ and $u_{r+1}^T y_i$ suffice. In Exercise 1.6, you are asked to show that this observation can be used to extend Matsui's Algorithm 1 to the case where only an estimate for $\Pr_{x_i} [u_1^T x_i = 0]$ is known (in addition to $u_{r+1}^T F(x)$). This observation is also used in Matsui's Algorithm 2, which is described in Section 1.4.2.

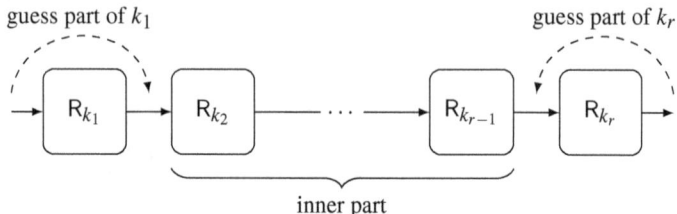

Figure 1.4 Partition of an iterated cipher into an inner and outer part.

### 1.4.2 Matsui's Algorithm 2

Matsui's Algorithm 2 divides the block cipher into two parts, as illustrated in Figure 1.4:

**The outer part,** where large parts of the round keys are recovered by guessing.

**The inner part,** where a linear approximation is used to filter the round key guesses made in the outer part, and where Matsui's Algorithm 1 can be applied to recover a linear expression in some bits of the expanded key.

For simplicity, we describe the algorithm for the case where the outer part consists of the last round only. More generally, it is possible to include multiple rounds at the beginning or end of the cipher, as long as not too many key bits have to be guessed.

If the inner part is denoted by $G_k$, then $E_k = R_{k_r} \circ G_k$, and hence $G_k = R_{k_r}^{-1} \circ E_k$. For a linear approximation $(u_1, u_r)$ of $G_k$, the empirical correlation is

$$\hat{\epsilon} = \frac{1}{q} \left| \left\{ 1 \leq i \leq q \mid u_1^T x_i = u_r^T R_{k_r}^{-1}(y_i) \right\} \right| - \frac{1}{2}.$$

Since $k_r$ is unknown a priori, we cannot determine $u_r^T R_{k_r}^{-1}(y)$ from $y$. However, a typical function $R_{k_r}$ has the property that for some masks $u_r$, a small number of bits of the round key $k_r$ are required to compute $u_r^T R_{k_r}^{-1}(y)$ from $y$.

Matsui's Algorithm 2 proceeds by estimating the empirical bias for every key guess, using the same random sample of plaintext-ciphertext pairs. It is assumed that for a wrong key guess, the empirical bias is close to zero – or at least much closer to zero than for the right value of the key. This assumption is often used in practice, although there are cases where a number of "equivalent" key guesses result in comparable empirical biases.

Matsui's Algorithm 2 outputs those key guesses for which the corresponding empirical bias is furthest away from zero. These remaining guesses are called candidate keys. Hence, Matsui's Algorithm 2 yields more information about the secret key than Matsui's Algorithm 1.

Determining the success rate of Matsui's Algorithm 2 would lead us too far at this point. For now we simply state that, like for Matsui's Algorithm 1, the required amount of data for a high success rate is proportional to $1/\epsilon_{u_1,u_r}^2$. In general, $\epsilon_{u_1,u_r} \geq \epsilon_{u_1,u_{r+1}}$. Hence, Matsui's Algorithm 2 might need less data than Matsui's Algorithm 1. However, the success rate of Matsui's Algorithm 2 also depends on the number of key values $K$ that are considered possible a priori and the number of candidate values that are returned as outputs. A more detailed analysis is given in Chapter 4.

When implemented naively, the $qK$ evaluations of $y \mapsto u_r^\mathsf{T} \mathsf{R}_{k_r}^{-1}(y)$ dominate the running time of Matsui's Algorithm 2. Later on, and in Chapter 5 in particular, faster ways to compute the empirical biases are discussed.

## 1.5 Remaining problems

At the end of this first chapter, it is worth pointing out some issues that have been ignored so far. By now, you are familiar with the basic idea of linear cryptanalysis. However, if you were to apply what you have learned to attack real block ciphers – or even the example cipher from Section 1.1 – you will likely run into trouble.

In Section 1.3, we used the piling-up lemma to estimate the bias of a linear approximation of a composition of functions. Understanding the accuracy of this estimate is an important part of Chapter 2.

Another issue is how to find the linear approximations and linear trails of a cipher with the highest bias (in absolute value). This problem is discussed in Chapter 3.

Finally, our discussion of key-recovery attacks in Section 1.4 ignores important aspects such as the success probability of the presented methods. This is important to understand precisely how much data is required to recover the key. This is discussed in detail in Chapter 4.

## 1.6 Historical remarks

Linear approximations and their bias are closely related to other concepts that were already used earlier on for the analysis of Boolean functions, such as the Walsh–Hadamard transformation and the minimum Hamming distance to an affine function. Exercise 1.5 explores the latter idea. Although these concepts were even explored in the context of cryptanalysis, the key ingredients of linear cryptanalysis were missing.

Linear approximations were first used in cryptanalysis by Anne Tardy-Corfdir and Henri Gilbert in 1991. They used linear approximations of parts of the block cipher FEAL to set up a key-recovery attack. The terms "linear cryptanalysis" and "piling-up lemma" were introduced by Matsui in 1993. He used linear cryptanalysis to attack the *Data Encryption Standard* (DES) block cipher.

## 1.7 References

Matsui, Mitsuru (May 1994a). "Linear Cryptanalysis Method for DES Cipher." In: *EUROCRYPT'93*. Ed. by Tor Helleseth. Vol. 765. LNCS. Springer, Berlin, Heidelberg, pp. 386–397. DOI: 10.1007/3-540-48285-7_33.

(Aug. 1994b). "The First Experimental Cryptanalysis of the Data Encryption Standard." In: *CRYPTO'94*. Ed. by Yvo Desmedt. Vol. 839. LNCS. Springer, Berlin, Heidelberg, pp. 1–11. DOI: 10.1007/3-540-48658-5_1.

Tardy-Corfdir, Anne and Henri Gilbert (Aug. 1992). "A Known Plaintext Attack of FEAL-4 and FEAL-6." In: *CRYPTO'91*. Ed. by Joan Feigenbaum. Vol. 576. LNCS. Springer, Berlin, Heidelberg, pp. 172–181. DOI: 10.1007/3-540-46766-1_12.

## 1.8 Exercises

### Exercise 1.1

Let $F(x) = k_2 + S(k_1 + x)$ with $k_1, k_2$ keys and S the S-box in Table 1.1.

1. Find a nontrivial effective linear approximation of F.
2. Apply Matsui's Algorithm 1 to recover one bit of the key.

Table 1.1. *A 4-bit S-box* S, *with values in hexadecimal (e.g.,* e $= 1110$*)*

| 0 | 1 | 2 | 3 | 4 | 5 | 6 | 7 | 8 | 9 | a | b | c | d | e | f |
|---|---|---|---|---|---|---|---|---|---|---|---|---|---|---|---|
| 8 | 2 | 4 | 0 | f | 5 | 7 | c | a | 6 | b | 3 | e | d | 9 | 1 |

### Exercise 1.2

A function $f: \mathbb{F}_2^n \to \mathbb{F}_2$ is called *linear* if $f(00\cdots 0) = 0$ and $f(x+y) = f(x) + f(y)$ for all $x$ and $y$ in $\mathbb{F}_2^n$.

For all $u$ in $\mathbb{F}_2^n$, let $\ell_u : \mathbb{F}_2^n \to \mathbb{F}_2$ be the linear function defined by $\ell_u(x) = u^T x$.

1. Show that for every linear function $f : \mathbb{F}_2^n \to \mathbb{F}_2$, there exists a mask $u$ in $\mathbb{F}_2^n$ such that $f = \ell_u$.
2. Give the truth tables of all 3-bit linear functions $\mathbb{F}_2^3 \to \mathbb{F}_2$.

### Exercise 1.3

Already before the introduction of linear cryptanalysis, it was known that the S-box $S_5 : \mathbb{F}_2^6 \to \mathbb{F}_2^4$ of DES (see Table 1.2) has a special "linear property."

1. Compute the LAT of $S_5$.
2. What is the special linear property?

Table 1.2. *Hexadecimal lookup table representation of the DES S-box $S_5$*

| 2 | e | c | b | 4 | 2 | 1 | c | 7 | 4 | a | 7 | b | d | 6 | 1 | ... |
|---|---|---|---|---|---|---|---|---|---|---|---|---|---|---|---|---|
| 8 | 5 | 5 | 0 | 3 | f | f | a | d | 3 | 0 | 9 | e | 8 | 9 | 6 | ... |
| 4 | b | 2 | 8 | 1 | c | b | 7 | a | 1 | d | e | 7 | 2 | 8 | d | ... |
| f | 6 | 9 | f | c | 0 | 5 | 9 | 6 | a | 3 | 4 | 0 | 5 | e | 3 |   |

### Exercise 1.4

Let $F : \mathbb{F}_2^n \to \mathbb{F}_2^{2n}$ be the *n*-bit forking operation defined by $F(x) = x \| x$. Compute the LAT of F.

### Exercise 1.5

Define the Hamming distance between functions $f : \mathbb{F}_2^n \to \mathbb{F}_2$ and $g : \mathbb{F}_2^n \to \mathbb{F}_2$ as $d_H(f, g) = \text{wt}(f + g)$, with wt the Hamming weight (number of ones) of the truth table of $f + g$.

1. The nonlinearity of a function $f : \mathbb{F}_2^n \to \mathbb{F}_2$ is defined by

$$\mathcal{N}(f) = \min_{\substack{u \in \mathbb{F}_2^n \\ a \in \mathbb{F}_2}} d_H(f, \ell_u + a).$$

Prove that $\mathcal{N}(f) = 2^{n-1} - \max_{u \in \mathbb{F}_2^n} \left| |\{x \in \mathbb{F}_2^n \mid f(x) = \ell_u(x)\}| - 2^{n-1} \right|$. See Exercise 1.2 for the notation $\ell_u$.

2. The nonlinearity of a function $F : \mathbb{F}_2^n \to \mathbb{F}_2^m$ is defined as

$$\mathcal{N}(F) = \min_{v \in \mathbb{F}_2^m \setminus \{0\}} \mathcal{N}(\ell_v \circ F).$$

Prove that
$$\mathcal{N}(\mathsf{F}) = 2^{n-1} - \max_{\substack{u \in \mathbb{F}_2^n \\ v \in \mathbb{F}_2^m \setminus \{0\}}} |\mathsf{LAT}_{u,v}^{\mathsf{F}}|.$$

### Exercise 1.6

1. Extend Matsui's Algorithm 1 to the case where only an estimate for the probability $\Pr_{\mathbf{x}}[u_1^\mathsf{T} \mathbf{x} = 0]$ is known in addition to $u_{r+1}^\mathsf{T} \mathsf{F}(\mathbf{x})$.
2. How would you use this extension if the plaintext is UTF-8-encoded English text?

### Exercise 1.7

Show that by choosing inputs carefully, the bias of a linear approximation of an invertible function can be computed by evaluating the function at only half of the inputs.

# 2
# Correlation matrices

In Chapter 1, we estimated the correlations of linear approximations by finding a suitable linear trail and applying the piling-up lemma – but this approach relied on an unjustified independence assumption. This chapter puts the piling-up lemma and linear cryptanalysis in general on a more solid theoretical foundation. This is achieved by using the theory of *correlation matrices*. Daemen proposed these matrices in 1994 to simplify the description of linear cryptanalysis.

## 2.1 Correlation of a random variable on $\mathbb{F}_2$

Recall from Chapter 1 that the bias of a linear approximation is the probability that it holds, minus one half. Throughout Chapter 1, and in Section 1.3 in particular, the term "bias" was also used more generally in relation to a random bit $\mathbf{x}$, i.e., a random variable on $\mathbb{F}_2$. Specifically, the bias of $\mathbf{x}$ is equal to

$$\epsilon_\mathbf{x} = \Pr_\mathbf{x}[\mathbf{x} = 0] - \frac{1}{2}.$$

In this chapter, it will be shown that the *correlation* of $\mathbf{x}$ is a more natural quantity to work with. The correlation of $\mathbf{x}$ is simply twice its bias:

$$c_\mathbf{x} = 2\epsilon_\mathbf{x} = 2\Pr_\mathbf{x}[\mathbf{x} = 0] - 1.$$

If $\mathbb{E}\mathbf{X}$ denotes the average of a random variable $\mathbf{X}$, then the correlation of a random bit $\mathbf{x}$ can also be written as

$$c_\mathbf{x} = \Pr_\mathbf{x}[\mathbf{x} = 0] - \Pr_\mathbf{x}[\mathbf{x} = 1] = \mathbb{E}(-1)^\mathbf{x},$$

since $(-1)^0 = 1$ and $(-1)^1 = -1$.

An ad hoc motivation to prefer correlations over biases is that they simplify the statement and proof of the piling-up lemma.

**Lemma 2.1** (Piling-up with correlations) *Let* $\mathbf{x}_1, \mathbf{x}_2, \ldots, \mathbf{x}_r$ *be random variables on* $\mathbb{F}_2$. *If* $\mathbf{x}_1, \ldots, \mathbf{x}_r$ *are independent, then the correlation of the sum* $\mathbf{x}_1 + \cdots + \mathbf{x}_r$ *satisfies*

$$c_{\mathbf{x}_1 + \cdots + \mathbf{x}_r} = \prod_{i=1}^{r} c_{\mathbf{x}_i} .$$

*Proof* If $\mathbf{X}$ and $\mathbf{Y}$ are independent random variables, then $\mathbb{E}\mathbf{XY} = (\mathbb{E}\mathbf{X})(\mathbb{E}\mathbf{Y})$. Hence,

$$c_{\mathbf{x}_1 + \cdots + \mathbf{x}_r} = \mathbb{E}\,(-1)^{\mathbf{x}_1 + \cdots + \mathbf{x}_r} = \mathbb{E}\,\textstyle\prod_{i=1}^{r}(-1)^{\mathbf{x}_i} = \prod_{i=1}^{r} \mathbb{E}\,(-1)^{\mathbf{x}_i} .$$

The second equality follows from the observation that $(-1)^{x+y} = (-1)^x(-1)^y$ for all $x$ and $y$ in $\mathbb{F}_2$, which can be checked case-by-case. □

Compared to Lemma 1.2 from Chapter 1, Lemma 2.1 does not involve additional factors of two. This is convenient, because it means that there is no need to keep track of the number of functions that are composed when applying the piling-up lemma in the context of linear cryptanalysis.

*Remark* 2.2 There is more to using correlations than notational convenience: the correlation $c_\mathbf{x}$ is the nontrivial coefficient of the Fourier transformation of the probability mass function of $\mathbf{x}$. As a matter of fact, the piling-up lemma is a special case of the convolution theorem for Fourier transformations. Later chapters will explain this link more clearly than is possible at present. ▷

## 2.2 Correlation between Boolean functions

In addition to the correlation of a random variable on $\mathbb{F}_2$, there is a related notion of correlation between two Boolean functions which is sometimes useful. Specifically, the correlation *between* functions $f \colon \mathbb{F}_2^n \to \mathbb{F}_2$ and $g \colon \mathbb{F}_2^n \to \mathbb{F}_2$ is defined as

$$C(f, g) = 2 \Pr_{\mathbf{x}}[f(\mathbf{x}) = g(\mathbf{x})] - 1,$$

with $\mathbf{x}$ uniform random on $\mathbb{F}_2^n$. This is nothing but the correlation of the random variable $f(\mathbf{x}) + g(\mathbf{x})$. Another name for $C(f, g)$ is the *correlation coefficient* between $f$ and $g$.

The correlation between two Boolean function is a measure of their similarity. If $f$ and $g$ are equal, then $C(f, g) = 1$. If $f$ and $g$ are always different, then $C(f,g) = -1$. Likewise, $C(f,g) = 0$ if $f$ and $g$ are equal on half of the inputs.

Using the properties of correlations of random variables, one can show that the correlation between Boolean functions $f$ and $g$ is equal to (see Exercise 2.1)

$$C(f,g) = \frac{1}{2^n} \sum_{x \in \mathbb{F}_2^n} (-1)^{f(x)+g(x)} = \left\langle (-1)^f, (-1)^g \right\rangle,$$

where $\langle \cdot, \cdot \rangle$ is an inner product between functions $\mathbb{F}_2^n \to \mathbb{R}$. This shows that correlations can be interpreted as inner products, which motivates the term "correlation."

Every linear Boolean function is of the form $\ell_u(x) = u^\mathsf{T} x$ for some $u$ in $\mathbb{F}_2^n$ (see Exercise 1.2). The correlation of a linear approximation $(u,v)$ with masks $u$ in $\mathbb{F}_2^n$ and $v$ in $\mathbb{F}_2^m$ of a function $\mathsf{F} \colon \mathbb{F}_2^n \to \mathbb{F}_2^m$ is equal to $C(\ell_u, \ell_v \circ \mathsf{F})$. This is indeed twice the bias $\epsilon_{u,v}$ of the approximation $(u,v)$.

## 2.3 Correlation matrices

In Section 1.2.2, the linear approximation table (LAT) of a function was defined. Definition 2.3 defines a similar table containing the correlations of all linear approximations of a function.

**Definition 2.3** (Correlation matrix) The correlation matrix of a function $\mathsf{F} \colon \mathbb{F}_2^n \to \mathbb{F}_2^m$ is a real $2^m \times 2^n$ matrix $C^\mathsf{F}$ with coordinates

$$C^\mathsf{F}_{v,u} = C(\ell_u, \ell_v \circ \mathsf{F}) = \frac{1}{2^n} \sum_{x \in \mathbb{F}_2^n} (-1)^{v^\mathsf{T} \mathsf{F}(x) + u^\mathsf{T} x}.$$

The correlation matrix of $\mathsf{F}$ is closely related to the LAT of $\mathsf{F}$:

$$\mathrm{LAT}_{u,v} = 2^{n-1} C^\mathsf{F}_{v,u}.$$

Mind the order of the indices: the output mask is the row-index for correlation matrices, but the column-index for the LAT. Unlike the LAT, the correlation matrix is more than a table containing the correlations of all linear approximations. It represents a linear operator between real vector spaces of dimensions $2^n$ and $2^m$. The properties of correlation matrices that follow below already suggest this, but a complete explanation will have to wait until Chapter 11. This has the downside that some of these properties may seem a bit miraculous, but this follows the historical development of the subject and allows keeping this chapter more concrete.

**Theorem 2.4** Let $\mathsf{F}\colon \mathbb{F}_2^n \to \mathbb{F}_2^m$ and $\mathsf{G}\colon \mathbb{F}_2^m \to \mathbb{F}_2^l$ be functions with correlation matrices $C^\mathsf{F}$ and $C^\mathsf{G}$, respectively. The correlation matrix of their composition $\mathsf{G}\circ\mathsf{F}$ is given by $C^{\mathsf{G}\circ\mathsf{F}} = C^\mathsf{G} C^\mathsf{F}$.

*Proof* By the definition of the product of two matrices, $\left(C^\mathsf{G} C^\mathsf{F}\right)_{v,u}$ is equal to

$$\sum_{w\in\mathbb{F}_2^m} C^\mathsf{G}_{v,w} C^\mathsf{F}_{w,u} = \sum_{w\in\mathbb{F}_2^m} \frac{1}{2^n}\frac{1}{2^m} \sum_{x\in\mathbb{F}_2^n}\sum_{y\in\mathbb{F}_2^m} (-1)^{v^\mathsf{T} \mathsf{G}(y)+w^\mathsf{T} y+w^\mathsf{T} \mathsf{F}(y)+u^\mathsf{T} x}$$

$$= \frac{1}{2^n} \sum_{x\in\mathbb{F}_2^n}\sum_{y\in\mathbb{F}_2^m} (-1)^{v^\mathsf{T} \mathsf{G}(y)+u^\mathsf{T} x} \frac{1}{2^m} \sum_{w\in\mathbb{F}_2^m} (-1)^{w^\mathsf{T}(y+\mathsf{F}(x))}.$$

The right-hand side can be rewritten as follows:

$$\sum_{w\in\mathbb{F}_2^m} C^\mathsf{G}_{v,w} C^\mathsf{F}_{w,u} = \frac{1}{2^n} \sum_{x\in\mathbb{F}_2^n}\sum_{y\in\mathbb{F}_2^m} (-1)^{v^\mathsf{T} \mathsf{G}(y)+u^\mathsf{T} x} \delta^y(\mathsf{F}(x))$$

$$= \frac{1}{2^n} \sum_{x\in\mathbb{F}_2^n} (-1)^{v^\mathsf{T} \mathsf{G}(\mathsf{F}(x))+u^\mathsf{T} x}.$$

In the first equation, $\delta^y\colon \mathbb{F}_2^m \to \mathbb{R}$ is the function defined by $\delta^y(y) = 1$ and $\delta^y(z) = 0$ for all $z \neq y$. The expression on the last line is precisely $C^{\mathsf{G}\circ\mathsf{F}}_{v,u}$. The first step uses the equality

$$\sum_{w\in\mathbb{F}_2^m} (-1)^{w^\mathsf{T} a} = \begin{cases} 2^m & \text{if } a = 0, \\ 0 & \text{otherwise}. \end{cases}$$

This follows from the fact that for all $t$ in $\mathbb{F}_2^m$,

$$\sum_{w\in\mathbb{F}_2^m} (-1)^{w^\mathsf{T} a} = \sum_{w\in\mathbb{F}_2^m} (-1)^{(w+t)^\mathsf{T} a} = (-1)^{t^\mathsf{T} a} \sum_{w\in\mathbb{F}_2^m} (-1)^{w^\mathsf{T} a}.$$

For all nonzero $a$, there exists at least one $t$ such that $t^\mathsf{T} a = 1$. Hence, the sum is its own opposite, which means it must be zero. If $a = 0$, then all terms in the sum are equal to one so that it equals $2^m$. □

Theorem 2.4 is the most important result related to correlation matrices. In theory, it provides a way to compute the correlations of linear approximations of $\mathsf{G}\circ\mathsf{F}$ given only the correlations of linear approximations of $\mathsf{F}$ and $\mathsf{G}$. In Chapter 1, the piling-up lemma was used to achieve this – but unlike the piling-up lemma, Theorem 2.4 does not require any independence heuristics.

Recall that an orthogonal matrix is a square matrix whose transpose is equal to its inverse. The following result shows that invertible functions have orthogonal correlation matrices.

## 2.4 Correlation matrices of structured functions

**Theorem 2.5** *Let* $F: \mathbb{F}_2^n \to \mathbb{F}_2^n$ *be a function with correlation matrix* $C^F$. *If* $F$ *is a permutation, then* $C^F$ *is an orthogonal matrix.*

*Proof* It is not difficult to see that the correlation matrix of the identity function is the identity matrix (this also follows from Theorem 2.6). Hence, it suffices to show that $C^{F^{-1}} = (C^F)^\mathsf{T}$. If $F$ is a permutation, then

$$C^F_{v,u} = \frac{1}{2^n} \sum_{x \in \mathbb{F}_2^n} (-1)^{v^\mathsf{T} F(x) + u^\mathsf{T} x} = \frac{1}{2^n} \sum_{y \in \mathbb{F}_2^n} (-1)^{v^\mathsf{T} y + u^\mathsf{T} F^{-1}(y)} = C^{F^{-1}}_{u,v}.$$

There is an alternative proof that works by computing $(C^F)^\mathsf{T} C^F$. □

*Example* 2.1 (Correlation matrix) The correlation matrix of the S-box $S: \mathbb{F}_2^3 \to \mathbb{F}_2^3$ of the example cipher from Section 1.1 is equal to

$$C^S = \begin{bmatrix} 1 & 0 & 0 & 0 & 0 & 0 & 0 & 0 \\ 0 & -1/2 & 0 & -1/2 & 0 & 1/2 & 0 & -1/2 \\ 0 & 0 & -1/2 & 1/2 & 0 & 0 & -1/2 & -1/2 \\ 0 & 1/2 & -1/2 & 0 & 0 & 1/2 & 1/2 & 0 \\ 0 & 0 & 0 & 0 & -1/2 & -1/2 & 1/2 & -1/2 \\ 0 & -1/2 & 0 & 1/2 & 1/2 & 0 & 1/2 & 0 \\ 0 & 0 & 1/2 & 1/2 & -1/2 & 1/2 & 0 & 0 \\ 0 & -1/2 & -1/2 & 0 & -1/2 & 0 & 0 & 1/2 \end{bmatrix}.$$

The coordinates of correlation matrices are indexed by bitvectors, so representing them as an array of numbers involves an arbitrary choice of ordering. The ordering chosen here is the lexicographic one: $001 \leq 010 \leq 011 \leq \cdots \leq 111$.

Since $S$ is a permutation, the matrix $C^S$ is orthogonal. Indeed, all its columns have Euclidean norm equal to one, and every pair of distinct columns is orthogonal:

$$\sum_{w \in \mathbb{F}_2^3} C^S_{w,u} C^S_{w,v} = \delta^u(v)$$

for all $u$ and $v$ in $\mathbb{F}_2^3$. The same property holds for every pair of rows. As in the proof of Theorem 2.4, $\delta^u: \mathbb{F}_2^n \to \mathbb{R}$ is the function defined by $\delta^u(u) = 1$ and $\delta^u(v) = 0$ for all $v \neq u$. ▷

## 2.4 Correlation matrices of structured functions

As discussed in Section 1.1, cryptographic functions are compositions of functions with special structure that makes it possible to evaluate them efficiently. This section gives the correlation matrix for two types of structured functions that are commonly used.

The first type are linear, and more generally affine, functions. The addition of the key and the bit permutation in the example cipher from Section 1.1 are examples of affine functions.

**Theorem 2.6** *Let* $\mathsf{F}\colon \mathbb{F}_2^n \to \mathbb{F}_2^m$ *be an affine map given by* $\mathsf{F}(x) = Ax + b$, *where $A$ is an $m \times n$ matrix over $\mathbb{F}_2$ and $b$ is a vector in $\mathbb{F}_2^m$. The correlation matrix $C^\mathsf{F}$ of $\mathsf{F}$ satisfies*

$$C^\mathsf{F}_{v,u} = (-1)^{v^\mathsf{T} b}\, \delta^u(A^\mathsf{T} v).$$

*Proof* The proof is by direct calculation:

$$C^\mathsf{F}_{v,u} = (-1)^{v^\mathsf{T} b} \frac{1}{2^n} \sum_{x \in \mathbb{F}_2^n} (-1)^{(A^\mathsf{T} v + u)^\mathsf{T} x} = (-1)^{v^\mathsf{T} b}\, \delta^u(A^\mathsf{T} v).$$

The second step follows from the equality $\sum_{x \in \mathbb{F}_2^n} (-1)^{w^\mathsf{T} x} = 2^n\, \delta^0(w)$, which was derived as a part of the proof of Theorem 2.4. $\square$

Theorem 2.6 can be interpreted as follows. For a given output mask, every linear function has only one possible effective linear approximation, and its input mask is a linear function of the given output mask. Furthermore, all the effective linear approximations have correlation one. Note that this is true even if the linear function is not invertible. For a constant addition, the input and output mask of an effective linear approximation must be equal, and the correlation is $\pm 1$ depending on the value of the constant.

The second and last class of structured functions discussed in this section are sometimes called "bricklayer maps." Figure 2.1 illustrates their structure.

A bricklayer map is a function $\mathsf{F}\colon \mathbb{F}_2^n \to \mathbb{F}_2^m$ built from $l$ functions $\mathsf{F}_1, \ldots, \mathsf{F}_l$ that operate on disjoint chunks of the input. More precisely,

$$\mathsf{F}(x_1 \| x_2 \| \cdots \| x_l) = \mathsf{F}_1(x_1) \| \mathsf{F}_2(x_2) \| \cdots \| \mathsf{F}_l(x_l),$$

where $\|$ denotes concatenation of bitvectors. The S-box layer of the example cipher from Section 1.1 is a good example.

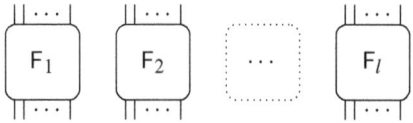

Figure 2.1 A bricklayer function built from $l$ functions $\mathsf{F}_1, \ldots, \mathsf{F}_l$.

## 2.4 Correlation matrices of structured functions

To describe the structure of correlation matrices of bricklayer maps, the Kronecker product of matrices is useful. Let $A$ be an $m \times n$ matrix over $\mathbb{R}$ with coordinates

$$A = \begin{bmatrix} A_{1,1} & A_{1,2} & \cdots & A_{1,n} \\ A_{2,1} & A_{2,2} & \cdots & A_{2,n} \\ \vdots & \vdots & \ddots & \vdots \\ A_{m,1} & A_{m,2} & \cdots & A_{m,n} \end{bmatrix}.$$

Similarly, let $B$ be a real $p \times q$ matrix. The Kronecker product of the matrices $A$ and $B$ is a real $pm \times qn$ matrix $A \otimes B$, equal to the block matrix

$$A \otimes B = \begin{bmatrix} A_{1,1}B & A_{1,2}B & \cdots & A_{1,n}B \\ A_{2,1}B & A_{2,2}B & \cdots & A_{2,n}B \\ \vdots & \vdots & \ddots & \vdots \\ A_{m,1}B & A_{m,2}B & \cdots & A_{m,n}B \end{bmatrix}.$$

Equivalently, the coordinates of $A \otimes B$ satisfy

$$(A \otimes B)_{p(i-1)+k, q(j-1)+l} = A_{i,j} B_{k,l}.$$

Alternatively, one could index the coordinates of $A \otimes B$ by pairs of indices. In this case, one has $(A \otimes B)_{(i,k),(j,l)} = A_{i,j} B_{k,l}$. As the coordinates of correlation matrices are indexed by bitvectors, the following convention is used:

$$\left(C^\mathsf{F} \otimes C^\mathsf{G}\right)_{v_1 \| v_2, u_1 \| u_2} = C^\mathsf{F}_{v_1, u_1} C^\mathsf{G}_{v_2, u_2},$$

with $C^\mathsf{F}$ and $C^\mathsf{G}$ the correlation matrices of functions $\mathsf{F}$ and $\mathsf{G}$, respectively.

**Theorem 2.7** Let $\mathsf{F}_1, \ldots, \mathsf{F}_l$ be functions with $\mathsf{F}_i \colon \mathbb{F}_2^{n_i} \to \mathbb{F}_2^{m_i}$, and let $n = \sum_{i=1}^{l} n_i$ and $m = \sum_{i=1}^{l} m_i$. The correlation matrix of the bricklayer function $\mathsf{F} \colon \mathbb{F}_2^n \to \mathbb{F}_2^m$ defined by $\mathsf{F}(x_1 \| \cdots \| x_l) = \mathsf{F}_1(x_1) \| \cdots \| \mathsf{F}_l(x_l)$ is equal to

$$C^\mathsf{F} = \bigotimes_{i=1}^{l} C^{\mathsf{F}_i}.$$

*Proof* Let $u = u_1\|\cdots\|u_l$ and $v = v_1\|\cdots\|v_l$. The coordinates of $C^\mathsf{F}$ satisfy

$$C^\mathsf{F}_{v,u} = \frac{1}{2^n} \sum_{x_1\|\cdots\|x_l \in \mathbb{F}_2^n} (-1)^{v_1^\mathsf{T} \mathsf{F}_1(x_1) + \cdots + v_l^\mathsf{T} \mathsf{F}_l(x_l) + u_1^\mathsf{T} x_1 + \cdots + u_l^\mathsf{T} x_l}$$

$$= \sum_{x_1 \in \mathbb{F}_2^{n_1}} \cdots \sum_{x_l \in \mathbb{F}_2^{n_l}} \prod_{i=1}^{l} \frac{1}{2^{n_i}} (-1)^{v_i^\mathsf{T} \mathsf{F}_i(x_i) + u_i^\mathsf{T} x_i}$$

$$= \prod_{i=1}^{l} \frac{1}{2^{n_i}} \sum_{x \in \mathbb{F}_2^{n_i}} (-1)^{v_i^\mathsf{T} \mathsf{F}_i(x_i) + u_i^\mathsf{T} x_i}$$

$$= \prod_{i=1}^{l} C^\mathsf{F}_{v_i, u_i}.$$

The result now follows by the definition of the Kronecker product. □

Another way to think about the proof of Theorem 2.7 is as an application of the piling-up lemma in the form of Lemma 2.1. This is acceptable because the inputs to the functions $\mathsf{F}_1, \ldots, \mathsf{F}_l$ are truly independent.

*Example* 2.2 (Key addition) Consider the addition of an $n$-bit key, i.e., the function $x \mapsto x + k$ on $\mathbb{F}_2^n$. For brevity, the correlation matrix of this function is denoted by $C^k$. One can think of this operation as a bricklayer map, since the $i$th bit of $x + k$ is just $x_i + k_i$. Hence, Theorem 2.7 implies that

$$C^k = \bigotimes_{i=1}^{n} \begin{bmatrix} 1 & 0 \\ 0 & (-1)^{k_i} \end{bmatrix}.$$

Alternatively, the same result can be obtained using Theorem 2.6.  ▷

## 2.5 Linear trails

Theorem 2.4 expresses the correlations of all linear approximations of a composition $\mathsf{F} = \mathsf{F}_r \circ \cdots \circ \mathsf{F}_1$ of $r$ functions in terms of the correlations of linear approximations of the $r$ functions $\mathsf{F}_1, \ldots, \mathsf{F}_r$ individually. In terms of correlation matrices,

$$C^\mathsf{F} = C^{\mathsf{F}_r} \cdots C^{\mathsf{F}_2} C^{\mathsf{F}_1}.$$

Although this is an interesting theoretical result, the large size of correlation matrices makes it difficult to use in practical calculations. To get around this issue, we exploit the sparsity of correlation matrices. In particular, writing out

## 2.5 Linear trails

the above product of matrices in terms of coordinates yields Corollary 2.8 below.

**Corollary 2.8** *Let* $F_1, \ldots, F_r$ *be functions on bitvectors. The correlation of a linear approximation of* $F = F_r \circ \cdots \circ F_1$ *is equal to the sum of the correlations of all linear trails with the same input and output mask as the approximation:*

$$C^F_{u_{r+1}, u_1} = \sum_{u_2, \ldots, u_r} \prod_{i=1}^{r} C^{F_i}_{u_{i+1}, u_i}.$$

If the matrices $C^{F_i}$ are sparse, then the sum in Corollary 2.8 contains a small number of nonzero terms. More generally, the idea is that a limited number of trails determines the value of the sum up to a small error. This is called the *dominant trail approximation*.

The traditional piling-up principle that was used in Chapter 1 is valid under the assumption that a single trail is dominant:

$$C^F_{u_{r+1}, u_1} \approx \prod_{i=1}^{r} C^{F_i}_{u_{i+1}, u_i}$$

for the trail $(u_1, u_2, \ldots, u_{r+1})$ with the largest absolute correlation. This explains why the piling-up lemma sometimes gives the correct result, even though the independence assumption it relies on does not hold.

For a key-alternating cipher $E_k = R_{k_r} \circ \cdots \circ R_{k_1}$ with $R_{k_i}(x) = R(x) + k_i$, Corollary 2.8 takes the form

$$C^{E_k}_{u_{r+1}, u_1} = \sum_{u_2, \ldots, u_r} (-1)^{\sum_{i=1}^{r} u_{i+1}^T k_i} \prod_{i=1}^{r} C^R_{u_{i+1}, u_i}. \tag{2.1}$$

In particular, the round keys influence the signs of trail correlations but not their absolute value.

*Example* 2.3 (Revisiting Example 1.3) Like in Example 1.3, consider the linear approximation $(000000001, 000010000)$ of three rounds of the example cipher from Section 1.1. In Example 1.3, a linear trail with the same input and output masks and correlation $(-1)^{k_0 + k_{10} + k_{22} + k_{31} + 1}/8$ was found. In light of Corollary 2.8, it is necessary to check whether or not there exist other trails that might affect the correlation of the linear approximation.

Since all bitvectors $u \ne 001$ such that $C^S_{u, 001} \ne 0$ contain at least two nonzero bits, at least two S-boxes in the second round of the cipher must have a nonzero input and output mask to obtain an effective trail. S-boxes with a nonzero output mask are called *active S-boxes*. Looking at the correlation matrix of the S-box layer in more detail, the only possibilities are

$u \in \{101, 011, 111\}$. Each of these choices leads to a unique effective trail. Figure 2.2 shows the resulting three effective trails.

The correlation of each trail can be calculated using Theorem 2.6 and 2.7. For example, consider the trail in Figure 2.2a. The correlation over the first round is equal to $(-1)^{k_0} C^S_{101,001} = (-1)^{k_0+1}/2$. For the second round, Theorem 2.7 implies that the correlation is

$$(-1)^{k_{10}+k_{16}} \underbrace{C^S_{010,010}}_{-1/2} \underbrace{C^S_{000,000}}_{1} \underbrace{C^S_{010,010}}_{-1/2} = (-1)^{k_{10}+k_{16}}/4.$$

Finally, the correlation over the third round is $(-1)^{k_{21}+k_{22}+k_{31}+1}/2$. Hence, the overall correlation of the trail is $(-1)^{k_0+k_{10}+k_{16}+k_{21}+k_{22}+k_{31}+1}/16$. A similar calculation for the other trails gives a total correlation of (check this!)

$$(-1)^{\kappa_1}/8 + (-1)^{\kappa_1+\kappa_2}/16 + (-1)^{\kappa_1+\kappa_3}/16 + (-1)^{\kappa_1+\kappa_2+\kappa_3}/32,$$

where $\kappa_1 = k_0 + k_{10} + k_{22} + k_{31} + 1$, $\kappa_2 = k_{16} + k_{21}$ and $\kappa_3 = k_{13} + k_{23}$. Since there are no other trails with the same input and output masks, the above expression is actually exact. Note that it can be rewritten as

$$(-1)^{\kappa_1}/8 \left(1 + (-1)^{\kappa_2}/2\right)\left(1 + (-1)^{\kappa_3}/2\right).$$

Based on the above, depending on the key, the correlation is either $\pm 1/32$, $\pm 3/32$ or $\pm 9/32$. In Example 1.4, a correlation close to $3/32$ was observed. This is because the key that was used in that example satisfies $\kappa_1 = 0$, $\kappa_2 = 1$ and $\kappa_3 = 0$.

▷

## 2.6 Historical remarks

Correlation matrices were introduced by Daemen in 1994 at the FSE conference, and in his PhD thesis from the following year. An important difference between Matsui's description of linear cryptanalysis and its description using correlation matrices is that Matsui assumed (round) keys to be uniform random variables. Moving away from random keys was an important advance that this book also follows, but one that was not fully appreciated at the time. Early work in linear cryptanalysis was often concerned with properties of key-averaged squared correlations. The most important example of such a result is the *linear hull theorem* due to Nyberg (see Exercise 2.11).

## 2.6 Historical remarks

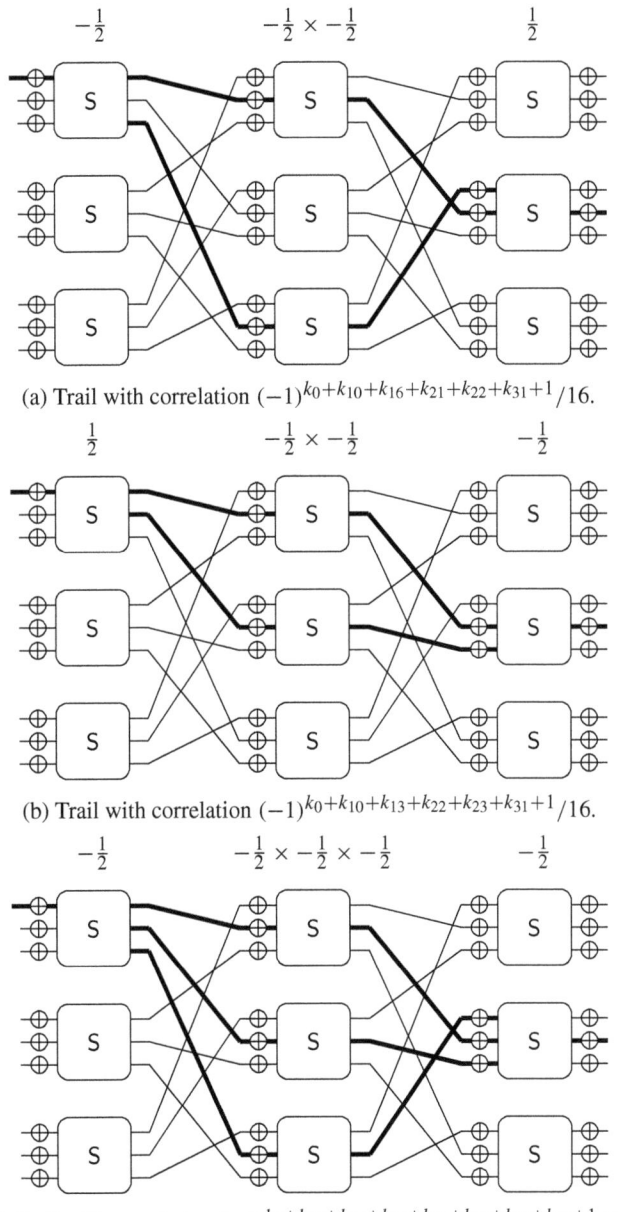

(a) Trail with correlation $(-1)^{k_0+k_{10}+k_{16}+k_{21}+k_{22}+k_{31}+1}/16$.

(b) Trail with correlation $(-1)^{k_0+k_{10}+k_{13}+k_{22}+k_{23}+k_{31}+1}/16$.

(c) Trail with correlation $(-1)^{k_0+k_{10}+k_{13}+k_{16}+k_{21}+k_{22}+k_{23}+k_{31}+1}/32$.

Figure 2.2 Three trails (a–c) with four or more active S-boxes.

## 2.7 References

Daemen, Joan (Mar. 1995). "Cipher and Hash Function Design Strategies Based on Linear and Differential Cryptanalysis." PhD thesis. KU Leuven.

Daemen, Joan, René Govaerts, and Joos Vandewalle (Dec. 1995). "Correlation Matrices." In: *FSE'94*. Ed. by Bart Preneel. Vol. 1008. LNCS. Springer, Berlin, Heidelberg, pp. 275–285. DOI: 10.1007/3-540-60590-8_21.

Nyberg, Kaisa (May 1995). "Linear Approximation of Block Ciphers (Rump Session)." In: *EUROCRYPT'94*. Ed. by Alfredo De Santis. Vol. 950. LNCS. Springer, Berlin, Heidelberg, pp. 439–444. DOI: 10.1007/BFb0053460.

## 2.8 Exercises

### Exercise 2.1

Let $f: \mathbb{F}_2^n \to \mathbb{F}_2$ and $g: \mathbb{F}_2^n \to \mathbb{F}_2$ be Boolean functions. Prove that

$$\mathsf{C}(f,g) = \langle (-1)^f, (-1)^g \rangle.$$

Here, $\langle \cdot, \cdot \rangle$ is the following inner product on the vector space of real-valued functions on $\mathbb{F}_2^n$:

$$\langle p, q \rangle = \frac{1}{2^n} \sum_{x \in \mathbb{F}_2^n} p(x) q(x),$$

for functions $p: \mathbb{F}_2^n \to \mathbb{R}$ and $q: \mathbb{F}_2^n \to \mathbb{R}$.

### Exercise 2.2

As shown in Exercise 1.2, every linear Boolean function on $\mathbb{F}_2^n$ is of the form $\ell_u$ with $u$ in $\mathbb{F}_2^n$ and $\ell_u(x) = u^\mathsf{T} x$.

1. Prove that for all $u$ and $v$ in $\mathbb{F}_2^n$, $\mathsf{C}(\ell_u, \ell_v) = \delta^u(v)$.
2. Use this property to rephrase the proof of Theorem 2.6 for linear functions.

### Exercise 2.3

Let $f: \mathbb{F}_2^n \to \mathbb{F}_2$ be a function. The Walsh–Hadamard transformation of $f$ is the function $\mathcal{W}_f: \mathbb{F}_2^n \to \mathbb{R}$ defined by

$$\mathcal{W}_f(u) = \frac{1}{2^n} \sum_{x \in \mathbb{F}_2^n} (-1)^{u^\mathsf{T} x + f(x)}.$$

Prove the following claims:
1. $W_f(u) = C(\ell_u, f)$.
2. $W_f(u) = 0$ if and only if $f + \ell_u$ is a balanced Boolean function. A Boolean function is balanced if it evaluates to zero on exactly half of its inputs.
3. The correlation of a linear approximation $(u, v)$ of $F: \mathbb{F}_2^n \to \mathbb{F}_2^m$ is $W_{\ell_v \circ F}(u)$.

## Exercise 2.4

This exercise explores a number of interesting consequences of Theorem 2.4 to 2.7. All of the following results follow from those theorems. Let $F: \mathbb{F}_2^n \to \mathbb{F}_2^n$ be a permutation. Prove the following claims:

1. For all $u$ in $\mathbb{F}_2^n$, the Walsh–Hadamard transformation (see Exercise 2.3) satisfies the identities $\sum_{v \in \mathbb{F}_2^n} W_{\ell_u \circ F}(v)^2 = 1$ and $\sum_{v \in \mathbb{F}_2^n} W_{\ell_v \circ F}(u)^2 = 1$. In particular, the first equality implies that $\sum_{v \in \mathbb{F}_2^n} W_f(v)^2 = 1$ for every Boolean function $f$. This is sometimes called Parseval's relation.
2. For all nonzero $v$ in $\mathbb{F}_2^n$, the Boolean function $\ell_v \circ F$ is balanced.

## * Exercise 2.5

1. Give an algorithm with time-complexity $\mathcal{O}_s(l\, s^l)$ to compute the matrix-vector product with a matrix $B$ of the form

$$B = A_1 \otimes \cdots \otimes A_l,$$

where $A_1, \ldots, A_l$ are $s \times s$ matrices and $l \geq 1$ is an integer.

2. Prove that for all $f: \mathbb{F}_2^n \to \mathbb{F}_2$ (see Exercise 2.3 for notation $\mathcal{W}_f$),

$$\begin{bmatrix} \mathcal{W}_f(0,0,\ldots,0) \\ \mathcal{W}_f(0,0,\ldots,1) \\ \vdots \\ \mathcal{W}_f(1,1,\ldots,1) \end{bmatrix} = \frac{1}{2^n} \left( \bigotimes_{i=1}^{n} \begin{bmatrix} 1 & 1 \\ 1 & -1 \end{bmatrix} \right) \begin{bmatrix} (-1)^{f(0,0,\ldots,0)} \\ (-1)^{f(0,0,\ldots,1)} \\ \vdots \\ (-1)^{f(1,1,\ldots,1)} \end{bmatrix}.$$

3. Deduce an algorithm to compute $\mathcal{W}_f$ in $\mathcal{O}(n2^n)$ time.
4. Deduce an algorithm to compute the correlation matrix of a given function $F: \mathbb{F}_2^n \to \mathbb{F}_2^m$ in $\mathcal{O}(n2^{n+m})$ time, assuming $F$ is given as a lookup table.

## Exercise 2.6

In this question you will analyze the construction in Figure 2.3. The input of the construction is denoted by $x$, the secret key by $k$. The correlation matrix of the S-box S is given in Example 2.1.

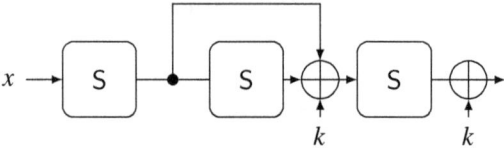

Figure 2.3 A construction with three S-boxes.

1. Find a linear trail with correlation $\pm 1/4$.
2. Find a linear approximation with correlation one for at least one key.
3. Suppose there exists an $x$ so that the corresponding output is 001. After learning this, and based on your answer to the previous question, what are the possible values of the key?

## Exercise 2.7

In this question you will analyze the construction $E_k : \mathbb{F}_2^6 \to \mathbb{F}_2^6$ in Figure 2.4. The secret key of the construction is denoted by $k = k_1 \| k_2$. The correlation matrix of the S-box S is given in Example 2.1.

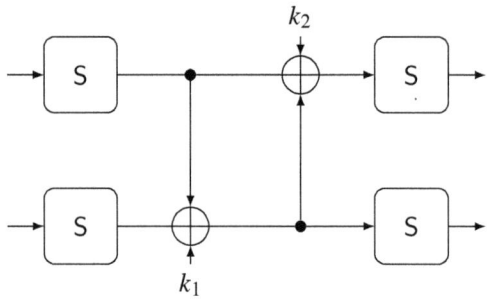

Figure 2.4 A construction with four S-boxes.

1. Find a linear trail with correlation $\pm 1/4$.
2. Find a nontrivial linear approximation with correlation one for $k = 000000$. Compute its correlation for all values of $k$.
3. Suppose that $E_k(000000) = 111111$. Based on your answer to the previous question, is it possible that $k_1 = k_2$?

### Exercise 2.8

Let $\text{and}_n : \mathbb{F}_2^{2n} \to \mathbb{F}_2^n$ denote the bitwise-and function on $n$-bit inputs:

$$\text{and}_n : (x_1, \ldots, x_n, y_1, \ldots, y_n) \mapsto (x_1 y_1, \ldots, x_n y_n).$$

This operation is used in some block ciphers (see Exercise 2.9). The next questions work towards a formula for the coordinates of the correlation matrix of $\text{and}_n$.

1. Show that $1 + (-1)^a + (-1)^b - (-1)^{a+b} = 2(-1)^{ab}$ for all $a$ and $b$ in $\mathbb{F}_2$.
2. Compute the correlation matrix of $\text{and}_1$ by hand.
3. Show that for all $n \geq 1$, the coordinates of the correlation matrix of $\text{and}_n$ are given by the formula

$$C^{\text{and}_n}_{w, u \| v} = \begin{cases} (-1)^{u^T v} / 2^{\text{wt}(w)} & \text{if } u \preccurlyeq w \text{ and } v \preccurlyeq w, \\ 0 & \text{otherwise}, \end{cases}$$

where $\text{wt}(w)$ is the number of nonzero coordinates of $w$ (often called the "Hamming weight" of $w$) and $\preccurlyeq$ denotes the ordering of bitstrings defined by $x \preccurlyeq y$ if and only if $x_i \leq y_i$ for all $i = 1, \ldots, n$.

### * Exercise 2.9

Simon is a so-called Feistel cipher designed by employees of the National Security Agency (NSA) of the United States. It is based on the bitwise-and function from Exercise 2.8. The round function of Simon-32 is illustrated in Figure 2.5. Bitwise-and is denoted by "∧," bitwise rotation by "⋘." For this exercise, make abstraction of how the round keys $k_i$ are generated.

1. Find a linear trail over two rounds with maximal absolute correlation.
2. Implement Simon-32 and estimate the correlation of the linear approximation you found in the previous question for the all-zero key. Do the results agree with what you expect? Try to explain potential discrepancies.

### * Exercise 2.10

A function $f : \mathbb{F}_2^n \to \mathbb{F}_2$ is called a *quadratic form* if it satisfies

$$f(x) = \sum_{i<j} a_{i,j} x_i x_j,$$

with $a_{i,j}$ elements of $\mathbb{F}_2$ and $x_1, \ldots, x_n$ the coordinates of $x$.

1. Prove that there exists an upper-triangular matrix $A$ so that $f(x) = x^T A x$.

2. Show that, up to a linear change-of-variables, the matrix $A$ can be taken as

3. Using Theorem 2.6 and the result of Exercise 2.8, derive a formula for the correlation matrix of an arbitrary quadratic form $f$ (this is a $2 \times 2^n$ matrix). Deduce that the absolute value of the correlation of every nontrivial linear approximation of $f$ is at most $2^{-r}$, with $r$ the rank of $A$.

4. Let $\mathsf{F}\colon \mathbb{F}_2^n \to \mathbb{F}_2^m$ be defined by $\mathsf{F}(x) = \mathsf{Q}(x) + \mathsf{L}(x)$ with $\mathsf{L}$ linear and so that every coordinate of $\mathsf{Q}$ is a quadratic form. Give an algorithm to compute the correlation of a given linear approximation of $\mathsf{F}$ in $\mathcal{O}(n^3)$ time.

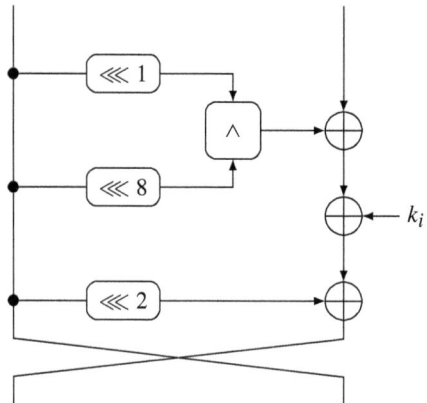

Figure 2.5 The round function of the block cipher Simon.

## Exercise 2.11

Consider a key-alternating cipher $E_k$ with round function R and independent and uniform random round keys $(k_1, \ldots, k_r) = k$. Prove the following claims:

1. $\mathbb{E}_k\left(C^{E_k}_{v,u}\right) = 0$ for all $(u,v) \neq (0,0)$.
2. For all $(u_{r+1}, u_1)$, the variance is

$$\mathbb{E}_k\left(C^{E_k}_{u_{r+1},u_1}\right)^2 = \sum_{u_2,\ldots,u_r} \prod_{i=1}^{r} \left(C^{R}_{u_{i+1},u_i}\right)^2.$$

This result was called the *linear hull theorem* by Nyberg.

# 3
# Optimization of linear trails

Finding linear trails with high absolute correlation quickly becomes tedious work, especially for ciphers with a more complicated structure than the example that we have worked with so far. Since the total number of trails is finite, finding linear trails with a maximal absolute correlation is an example of a combinatorial optimization problem.

This chapter discusses three commonly used optimization methods: Matsui's branch and bound method, mixed-integer linear programming and satisfiability or satisfiability modulo theories. At the same time, the chapter introduces two additional example ciphers that follow a different design strategy.

## 3.1 Branch and bound

The first optimization method that we discuss is due to Matsui. It is an example of a depth-first branch and bound search algorithm, but one can think of numerous variations on the basic strategy that might be advantageous.

### 3.1.1 Depth-first search

A depth-first algorithm traverses the vertices of a graph by following its edges as far out as possible before backtracking. Figure 3.1 illustrates this process for the case where the graph is a tree. Starting at the root $a$, in each step the algorithm selects one of the children of the previously selected vertex. In Figure 3.1, the vertices $a$, $b$, $c$ and $d$ are visited successively. Vertex $d$ does not have any children, so the algorithm jumps back to the last vertex with unexplored children. This is vertex $b$, and from here the search continues with $e$ and $f$. Since $f$ does not have any children, the algorithm jumps back to

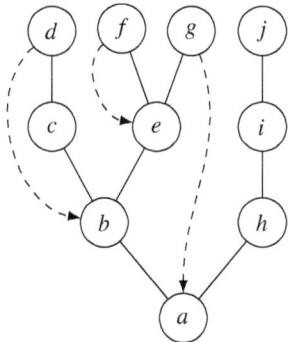

Figure 3.1 A depth-first traversal of a tree of height three.

vertex $e$. After visiting $g$, it jumps back to the root $a$ and successively explores $h$, $i$ and $j$.

Finding shortest paths in an edge-weighted graph is a typical application of depth-first algorithms. A shortest path between two vertices is a sequence of connected vertices that minimizes the sum of the weights on the edges between successive vertices. More generally, one can consider shortest paths between two sets of vertices. For example, the depth-first traversal in Figure 3.1 could have been used to find a shortest path between the root $a$ and some leaf $d$, $f$, $g$ or $j$ by keeping track of the length of the path followed by the algorithm. To find a shortest path, this method visits all vertices at least once. However, one can hope to find a reasonably short path early on by using heuristics to select the next child to visit. One strategy is to consistently follow the lowest-weight edge first. This is known as greedy search.

Branch and bound is a method to find a shortest path without visiting all vertices. It relies on the assumption that extending a path never decreases its total weight. If all weights are nonnegative, this is guaranteed to be true. Branch and bound can be combined with depth-first search by modifying the selection of the next child vertex as follows. Let $x$ be the current vertex, $f(x)$ the total weight of the current path from the root to $x$ and $B$ the total weight of the best path found so far. The algorithm also uses a lower bound $h(y)$ on the total weight of a shortest path starting at a vertex $y$. The function $h\colon V \to \mathbb{R}$ is called the heuristic. A child $y$ of $x$ can be ignored whenever

$$f(x) + w_{x,y} + h(y) \geq B,$$

where $w_{x,y}$ is the weight on the edge connecting $x$ and $y$. Indeed, $f(x) + w_{x,y} + h(y)$ is a lower bound on the total weight of all paths that might be found by continuing the search from $y$.

The number of vertices visited by a depth-first branch and bound algorithm heavily depends on the heuristic function $h$. In the worst case, $h(y) = 0$ for all vertices $y$.

### 3.1.2 Matsui's method

Let $F = F_r \circ \cdots \circ F_1$ be a composition of $r$ functions $F_1, \ldots, F_r$ with $F_i \colon \mathbb{F}_2^{n_i} \to \mathbb{F}_2^{n_{i+1}}$. Recall that a trail is a sequence $(u_1, \ldots, u_{r+1})$ of masks. The set of all possible trails forms a graph $G$ with vertex set

$$V = \bigcup_{1 \le i \le r+1} \{(i, u_i) \mid u_i \in \mathbb{F}_2^{n_i}\}.$$

The edges are between vertices $(i, u_i)$ and $(i+1, u_{i+1})$ with $C^{F_i}_{u_{i+1}, u_i} \ne 0$, and their weight is equal to $-\log_2 |C^{F_i}_{u_{i+1}, u_i}|$. If the edges have an orientation (from $i+1$ to $i$), then $G$ is a directed acyclic graph. The total weight of a path $(1, u_1), (2, u_2), \ldots, (r+1, u_{r+1})$ is equal to

$$\sum_{i=1}^{r} -\log_2 \left|C^{F_i}_{u_{i+1}, u_i}\right| = -\log_2 \prod_{i=1}^{r} \left|C^{F_i}_{u_{i+1}, u_i}\right|.$$

Hence, a shortest path between vertices of the form $(1, u_1)$ and $(r+1, u_{r+1})$ corresponds to a trail with maximal absolute correlation. Matsui's method finds such trails in a depth-first manner, starting in the last round and working towards the first round.[1] The method is summarized in Algorithm 3.1.

The heuristic function $h \colon V \to \mathbb{R}$ used by Algorithm 3.1 is specified in terms of bounds $B_1, \ldots, B_r$ on the maximum absolute correlation of a trail from round one to round $i$. That is,

$$h\bigl((i, u_i)\bigr) = -\log_2 B_i.$$

In practice, the heuristic function can often be improved by taking into account the specific value of the mask $u_i$. Note that one can always choose $B_i = 1$.

The order in which the next masks are enumerated (lines 10 and 17) is left unspecified in Algorithm 3.1, but it is important in practice. For the inner loop (line 10), a greedy approach is usually used: choose the masks $u_{l-1}$ in order of decreasing $|C^{F_i}_{u_l, u_{l-1}}|$. However, more variation is possible in the way ties are broken. For the outer loop (line 17), another approach is necessary. Typically, one uses a heuristic based on looking ahead by one round. For example, one can count the number of S-boxes that are guaranteed to be active in the last round.

---

[1] Matsui's original method starts with the first round, but working backwards is better when some of the functions $F_1, \ldots, F_r$ are not invertible.

## 3.1 Branch and bound

**Algorithm 3.1** Outline of Matsui's method for finding trails.

**Input:**
   Functions $F_1, \ldots, F_r$ with correlation matrices $C^{F_1}, \ldots, C^{F_r}$
   Bounds $B_0 = 1, B_1, \ldots, B_r$ such that $B_l \geq \prod_{i=1}^{l-1} \left| C_{u_{i+1}, u_i}^{F_i} \right|$ for all $u_1, \ldots, u_l$

**Output:**
   Trail $(v_1, \ldots, v_{r+1})$ with maximal $\prod_{i=1}^{r} \left| C_{v_{i+1}, v_i}^{F_i} \right|$

1: ▷ Initialize absolute correlation of the best trail found so far:
2: $B \leftarrow 0$
3: ▷ Recursive procedure to find the best trail, starting at the last round
4: **procedure** SEARCH$(u_{r+1}, u_r, \ldots, u_l)$
5:    **if** $l = 1$ **then**
6:       $B \leftarrow \prod_{i=1}^{r} \left| C_{u_{i+1}, u_i}^{F_i} \right|$
7:       $(v_1, \ldots, v_{r+1}) \leftarrow (u_1, \ldots, u_{r+1})$
8:       **return**
9:    **end if**
10:    **for** all $u_{l-1}$ such that $C_{u_l, u_{l-1}}^{F_l} \neq 0$ **do** ▷ Heuristics to determine order
11:       **if** $B_{l-1} \prod_{i=l-1}^{r} \left| C_{u_{i+1}, u_i}^{F_i} \right| > B$ **then**
12:          SEARCH $(u_{r+1}, u_r, \ldots, u_l, u_{l-1})$
13:       **end if**
14:    **end for**
15: **end procedure**
16:
17: **for** all $u_{r+1} \neq 0$ **do**                     ▷ Heuristics to determine order
18:    SEARCH $(u_{r+1})$
19: **end for**
20:
21: **return** $(v_1, \ldots, v_{r+1})$

*Example* 3.1 Let us apply Algorithm 3.1 to three rounds of the example cipher from Section 1.1. Choose $B_1 = 1$. The absolute correlation of an approximation over one round is at most $1/2$, since at least one S-box must be active. Hence, $B_2 = 1/2$ and $B_3 = 1/4$ are admissible choices.

Let $u_1$, $u_2$ and $u_3$ denote the masks at the input of the first three rounds of the cipher, respectively. Choose the output mask $u_4$ to minimize the number of active S-boxes in the last round, breaking ties at random. For example, this could lead to the choice $u_4 = 000010000$. Continuing with this choice suggests

$\{000010000, 000011000, 000101000, 000111000\}$ as candidate values for $u_3$. Breaking ties at random again, suppose that 000011000 is selected. For $u_2$, possible choices are then

$$\{010000010, 011000010, 101000010, 111000010, 010000011, \ldots\}.$$

Again, suppose breaking ties at random leads to the choice 011000010. At this point, all valid choices of $u_1$ will result in a trail with absolute correlation $1/16$. Having found such a trail, the algorithm returns to the selection of $u_2$. However, since $B_2/8 = 1/16$, no other choice of $u_2$ can result in a better trail. As a result, the algorithm immediately backtracks to the choice of $u_3$. Since $B_3/2 = 1/8$, all other choices of $u_3$ are still valid candidates. However, for every choice other than 000010000, the algorithm immediately backtracks.

After selecting 000010000 as the value of $u_3$, the only choice of $u_2$ that is not ruled out as a dead end is 000000010. Every valid choice of $u_1$ then gives a trail with correlation $1/8$, which is optimal. The algorithm will immediately backtrack to the choice of $u_4$. By aborting early based on the number of S-boxes that will be active in the last round, the algorithm can terminate after checking $3 \cdot 8 - 2 = 22$ more values of $u_4$. ▷

Matsui's branch and bound method is also applicable to other types of ciphers, including the examples that are introduced in Sections 3.2.1 and 3.3.1. However, this is tedious and error-prone work – especially when efficiency becomes important. An alternative approach is to reformulate the optimization problem as a mixed-integer linear programming (Section 3.2) or satisfiability (Section 3.3) problem, so that it can be solved using off-the-shelf software.

## 3.2 Mixed-integer linear programming

A linear programming problem is an optimization problem in real variables $x_1, \ldots, x_n$, with a linear objective function:

$$\min_{x_1, \ldots, x_n} \sum_{i=1}^{n} a_i x_i,$$

where $a_1, \ldots, a_n$ are given real numbers. The variables $x_1, \ldots, x_n$ are constrained by an arbitrary number of linear inequalities of the form

$$\sum_{i=1}^{n} b_i x_i \geq b_{n+1},$$

## 3.2 Mixed-integer linear programming

with $b_1, \ldots, b_n$ and $b_{n+1}$ given real numbers. Linear programming problems can be solved efficiently in practice, e.g., using the simplex algorithm. If some of the variables $x_1, \ldots, x_n$ are required to be integers, the problem is called a mixed-integer linear programming problem. For applications in linear cryptanalysis, the integer variables are usually restricted to be $\{0, 1\}$-valued. In general, linear programming problems with integer variables are much more difficult to solve. Nevertheless, there exists specialized software that can tackle such problems in practice. This section takes the existence of such "solvers" for granted and instead focuses on how to express the optimization of trails as a mixed-integer linear programming problem.

### 3.2.1 Example: Rijndael-like cipher

This section introduces a second example cipher that will be analyzed using mixed-integer linear programming in Sections 3.2.2 and 3.2.3. It is a key-alternating cipher that follows the *wide-trail design strategy* to withstand linear cryptanalysis.

The overall structure of the round function is shown in Figure 3.2. It is a function on $\mathbb{F}_2^{96}$, but the description of the round function is more convenient when the bits of the state are laid out in a $4 \times 8$ grid of 3-bit groups ("cells"). The first operation is a bricklayer function or S-box layer and is called SubCells. The second and third steps are both linear functions: the second step rotates the rows of the state and is called ShiftRows, and the third step applies a linear function to every column and is called MixColumns. Hence, the unkeyed round function is

$$R = \text{MixColumns} \circ \text{ShiftRows} \circ \text{SubCells}.$$

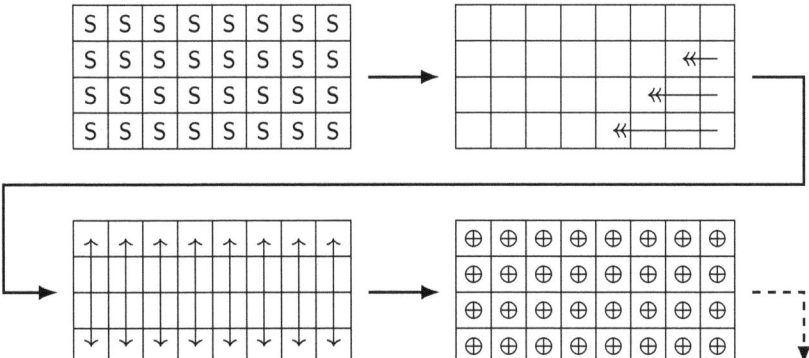

Figure 3.2 The round function consists of SubCells, ShiftRows, MixColumns and round key addition. Source: Beierle et al. (2018). © IACR.

Additional details about the functions on the right-hand side are given below.

**SubCells** consists of the parallel application of an S-box S to the 3-bit cells of the state. The S-box $S: \mathbb{F}_2^3 \to \mathbb{F}_2^3$ is the same as the S-box of the example cipher from Section 1.1.

**ShiftRows** shuffles the cells of the state. If the rows are numbered from zero to three with zero corresponding to the top row, then ShiftRows rotates the $i$th row of the state over $3 \cdot i$ bits to the left.

**MixColumns** applies a linear map to every column of the state. Let $(x_1, \ldots, x_4)$ with $x_i$ in $\mathbb{F}_2^3$ denote a column of the state. MixColumns maps $(x_1, \ldots, x_4)$ to a new column given by (with $I$ the $3 \times 3$ identity matrix)

$$\begin{bmatrix} 0 & I & I & I \\ I & 0 & I & I \\ I & I & 0 & I \\ I & I & I & 0 \end{bmatrix} \begin{bmatrix} x_1 \\ x_2 \\ x_3 \\ x_4 \end{bmatrix}.$$

Exercise 3.1 asks you to show that this map is invertible.

The $i$th round function is defined as $\mathsf{R}_{k_i}(x) = \mathsf{R}(x) + k_i$, and the cipher is the composition of $\mathsf{R}_{k_1}, \ldots, \mathsf{R}_{k_r}$ preceded by an initial round key addition with key $k_0$. The number of rounds $r$ will be left open for now. The round keys are $k_i = k + c_i$ for $i = 0, \ldots, r$, with $k$ in $\mathbb{F}_2^{96}$ the key of the cipher and $c_1, \ldots, c_r$ constants. The values of the constants on the state cells (from left to right and top to bottom, from the first to the last round) are given by the Thue–Morse sequence $000, 111, 111, 000, 111, 000, 000, 111, \ldots$

The linear layer, and the MixColumns step in particular, is chosen to ensure that every linear trail involves a high number of active S-boxes. This is an important difference with the example cipher of Section 1.1, for which there are linear trails with only one active S-box per round. By Theorem 2.6, the effective linear approximations over a linear function $x \mapsto Mx$ with $M$ in $\mathbb{F}_2^{n \times n}$ are all of the form $(M^\mathsf{T} u, u)$ with $u$ in $\mathbb{F}_2^n$. Hence, it makes sense to choose the matrix $M$ so that the total number of nonzero cells in $u$ and $M^\mathsf{T} u$ is large. Indeed, these nonzero cells correspond to active S-boxes in the rounds before and after the linear layer. This motivates the following definition.

**Definition 3.1** (Linear branch number) Let $M$ be a $bm \times bn$ matrix over $\mathbb{F}_2$. The linear branch number of $M$ (or the function $x \mapsto Mx$) is equal to

$$\mathcal{B}(M) = \min_{x \neq 0} \mathsf{wt}_m(x) + \mathsf{wt}_n(M^\mathsf{T} x),$$

where $\mathsf{wt}_l(x_1 \| x_2 \| \cdots \| x_l) = |\{1 \leq i \leq l \mid x_i \neq 0\}|$ is the Hamming weight.

## 3.2 Mixed-integer linear programming

In Exercise 3.1, you will show that the branch number of MixColumns is equal to four. This means that a linear trail over two rounds always involves at least four active S-boxes. In Exercise 3.3, you will show that every four-round trail involves at least 16 active S-boxes.

### 3.2.2 Building a model

To formulate the optimization problem as a mixed-integer linear program, introduce binary integer variables corresponding to the bits of the input and output masks of every S-box layer. Modeling the ShiftRows step is straightforward, as it amounts to shuffling (relabeling) the variables. The modeling of MixColumns and SubCells is discussed below.

The MixColumns step is based on exclusive-or operations. The condition that $z$ is the exclusive-or of $x$ and $y$ can be expressed using linear inequalities in more than one way. For example, assuming that $x$, $y$ and $z$ are binary variables,

$$x + y + z \leq 2,$$
$$x + y + z \geq 2d,$$
$$d \geq x,$$
$$d \geq y,$$
$$d \geq z,$$

where the additions are in the integers and $d$ is a new integer dummy variable. If $d$ is a binary variable, then the following linear equality also works:

$$x + y + z = 2d.$$

To implement $M$, the exclusive-or between three variables is needed. To express that $w$ is the exclusive-or of $x$, $y$ and $z$, the above can be generalized to

$$w + x + y + z = 4d_1 - 2d_2 - 2d_3$$

for binary dummy variables $d_1$, $d_2$ and $d_3$. In particular, if $(v_1, v_2, v_3, v_4)$ is an output mask for $x \mapsto Mx$ and $(u_1, u_2, u_3, u_4)$ the input mask, then

$$u_{1,i} + v_{2,i} + v_{3,i} + v_{4,i} = 4d_1 - 2d_2 - 2d_3,$$
$$u_{2,i} + v_{1,i} + v_{3,i} + v_{4,i} = 4e_1 - 2e_2 - 2e_3,$$
$$u_{3,i} + v_{1,i} + v_{2,i} + v_{4,i} = 4f_1 - 2f_2 - 2f_3,$$
$$u_{4,i} + v_{1,i} + v_{2,i} + v_{3,i} = 4g_1 - 2g_2 - 2g_3,$$

where $u_{1,i}$ is the $i$th bit of $u_1$ and similarly for $u_2, u_3, u_4$ and $v_1, \ldots, v_4$.

To model SubCells, introduce a new variable for each S-box to indicate whether or not it is active. In addition, ineffective linear approximations over the S-box must be ruled out. Assume that $(u_1, u_2, u_3)$ is the input mask and $(v_1, v_2, v_3)$ the output mask. If $a$ is a binary variable that is one when the S-box is active and zero otherwise, then

$$3a \geq v_1 + v_2 + v_3,$$

where the addition is again over the integers. Ruling out linear approximations over the S-box with correlation zero is more difficult. However, there is a general approach to this problem. Recall that a set $S \subset \mathbb{R}^n$ is called convex if, for all $x$ and $y$ in $S$, all points on the connecting line $\{\lambda x + (1-\lambda)y \mid \lambda \in [0,1]\}$ are also contained within $S$. The convex hull of a set $T \subseteq \mathbb{R}^n$ is the smallest convex set that includes $T$.

The convex hull of a finite set is a convex polytope. For example, the convex hull of the set $\{(0,0), (1,0), (0,1)\} \subset \mathbb{R}^2$ is shown in Figure 3.3. Convex polytopes have the useful property that they can be described by a finite number of linear inequalities. This is because every linear inequality cuts out a half-plane.

Hence, a set of inequalities that rules out ineffective linear approximations over S can be found by finding the linear inequalities that cut out the convex hull of the set (up to the conventional map $\mathbb{F}_2 \hookrightarrow \{0,1\} \subset \mathbb{R}$)

$$T = \left\{ (u_1, u_2, u_3, v_1, v_2, v_3) \in \mathbb{F}_2^6 \mid C^{\mathsf{S}}_{v_1 \| v_2 \| v_3, u_1 \| u_2 \| u_3} \neq 0 \right\} \subset \mathbb{R}^6.$$

Importantly, the convex hull of $T$ does not contain any points in $\{0,1\}^6$ other than those in $T$. Without going into details, the convex hull of a set can be computed using the *Quickhull algorithm*. The set of linear inequalities obtained from the convex hull representation is not necessarily minimal. Hence, additional methods are often used to reduce the number of linear inequalities. Finding a minimal subset of inequalities is a set cover problem and can itself be solved using mixed-integer linear programming.

Figure 3.3 The convex hull of the set $\{(0,0), (1,0), (0,1)\}$.

## 3.2 Mixed-integer linear programming

If $a_1, \ldots, a_l$ are the binary variables indicating whether or not the S-boxes are active, then the objective function is equal to

$$\sum_{i=1}^{l} a_i .$$

This is equal to the weight of the linear trail expressed by all of the variables. This assumes that the nonzero correlations of all linear approximations over the S-box are equal to $\pm 1/2$. In general, it may be necessary to encode the weights as integer variables or to define multiple variables corresponding to linear approximations with different weights.

### 3.2.3 Solving the model

Popular MILP solvers include CPLEX and Gurobi. Problems are either submitted to the solver programmatically using a solver-specific API, or given as a file. Most solvers support the LP format. LP files are text files that contain the objective function, the inequalities and the type of all variables. For example:

```
\ Objective (minimize or maximize)
minimize
x + y + z
\ List of inequalities
subject to
2 - x - y - z >= 0
x + y + z - 2d >= 0
d - x >= 0
d - y >= 0
d - z >= 0
\ Variable types
generals
d
binary
x y z
end
```

## 3.3 Satisfiability and satisfiability modulo theories

Satisfiability or "SAT" is a decision problem that asks whether or not there exists an assignment of variables so that a given Boolean formula is true. In practice, it is often the search variant of this problem that is useful – but a valid assignment can be found efficiently by repeatedly solving the decision problem. Most solvers assume that the Boolean formula is in *conjunctive normal form*. Specifically, every Boolean formula in $n$ variables $x_1, \ldots, x_n$ in $\mathbb{F}_2$ can be written as

$$\bigwedge_{i=1}^{m} (x_{i_1} + b_{i_1}) \vee (x_{i_2} + b_{i_2}) \vee \cdots \vee (x_{i_{l_i}} + b_{i_{l_i}}),$$

where $\wedge$ denotes "and," $\vee$ denotes "or," and $b_{i_1}, \ldots, b_{i_{l_i}}$ are constants in $\mathbb{F}_2$.

Satisfiability modulo theories (SMT) extend the satisfiability problem to more general formulas that can include quantifiers, integer variables, bitvectors, ... Internally, SMT solvers often convert at least part of the problem to a SAT instance.

Both SAT and SMT are difficult to solve in general. As in the case of mixed-integer linear programming, we take the existence of solvers that can deal with practical instances for granted.

### 3.3.1 Example: Add-rotate-xor cipher

Some ciphers do not rely on small S-boxes to introduce nonlinearity, but on operations with long operands that are natively supported by modern processors (such as modular addition or bitwise-and). The example given in this section is part of the add-rotate-xor (ARX) family of designs. The optimization of trails in such ciphers can often be conveniently modeled as an SMT problem with bitvector variables.

The round function of the example is shown in Figure 3.4b. The symbol "⊞" represents integer addition modulo $2^n$, for $n$ in $\{24, 32, 48, 64\}$. The round keys are generated by iterating a similar function, as shown in Figure 3.4a. This cipher is called Speck and was designed by employees of the National Security Agency of the United States.

Similar to the Rijndael-like cipher from Section 3.2.1, the number of rounds will be left open for now. Unlike for the Rijndael-like cipher, there is no simple argument to upper bound the absolute correlation of trails for Speck.

### 3.3.2 Building a model

The coordinates of the correlation matrices of the linear functions in the cipher (rotations and exclusive-or between two branches) have closed-form formulas

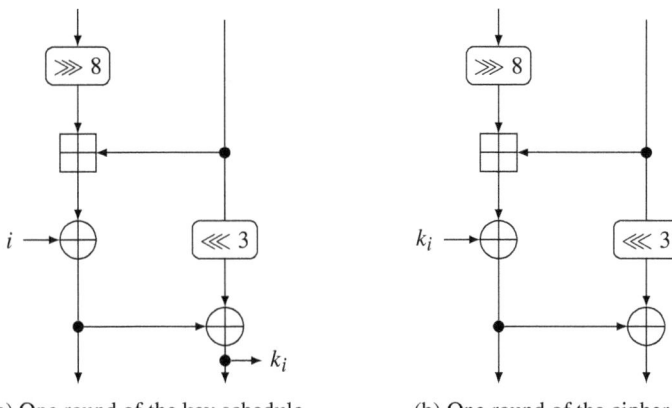

(a) One round of the key schedule.   (b) One round of the cipher.

Figure 3.4 The ARX cipher Speck.

that can be converted into bitvector constraints in a straightforward way. The addition modulo a power of two poses the main difficulty. However, it turns out that there is also a simple formula for the coordinates of the correlation matrix of this operation.

The formula is derived by transforming the graph of the modular addition function to the graph of a function that is easier to deal with. The graph of a function $\mathsf{F}\colon \mathbb{F}_2^n \to \mathbb{F}_2^m$ is the set of pairs $G_\mathsf{F} = \{(x, \mathsf{F}(x)) \mid x \in \mathbb{F}_2^n\}$. In particular, $G_\boxplus$ is the graph of the modular addition function $x \parallel y \mapsto x \boxplus y$ with modulus $2^n$.

**Lemma 3.2** (Schulte–Geers) *Let $M \in \mathbb{F}_2^{n \times n}$ be the lower-triangular matrix*

$$M = \begin{bmatrix} 0 & 0 & 0 & \cdots & 0 & 0 \\ 1 & 0 & 0 & \cdots & 0 & 0 \\ 1 & 1 & 0 & \cdots & 0 & 0 \\ \vdots & \vdots & \vdots & \ddots & \vdots & \vdots \\ 1 & 1 & 1 & \cdots & 0 & 0 \\ 1 & 1 & 1 & \cdots & 1 & 0 \end{bmatrix}.$$

*Furthermore, let $\mathsf{Q}\colon \mathbb{F}_2^{2n} \to \mathbb{F}_2^n$ be the function defined by $\mathsf{Q}(x \parallel y) = M(x \wedge y)$, where $x \wedge y$ is the bitwise-and of $x$ and $y$. The map $(x \parallel y, z) \mapsto ((x + z) \parallel (y + z), x + y + z)$ is a bijection from $G_\boxplus$ to $G_\mathsf{Q}$.*

*Proof* It is not difficult to see that $(x \parallel y, z) \mapsto ((x + z) \parallel (y + z), x + y + z)$ is a bijection if it is well-defined. Hence, it suffices to verify that if $z = x \boxplus y$, then

$$x + y + z = \mathsf{Q}(x + z \parallel y + z).$$

However, $c = x + y + z$ is equal to the vector of carry bits obtained during the modular addition of $x$ and $y$. These carry bits $c$ satisfy $c_1 = 0$ and

$$c_{i+1} = c_i + (x_i + z_i)(y_i + z_i).$$

The above relation follows from the grade school algorithm for addition. Hence, the carry vector $c$ is equal to $M\big((x+z) \wedge (y+z)\big) = \mathsf{Q}(x+z \,\|\, y+z)$. □

In Theorem 3.3, the notation $\preccurlyeq$ refers to the coordinate-wise ordering of bitvectors of length $n$ defined by $0 \preccurlyeq 0$, $0 \preccurlyeq 1$ and $1 \preccurlyeq 1$. Note that $x \preccurlyeq y$ is the same as "$x_i$ implies $y_i$" for $i = 1, \ldots, n$. Hence, $x \preccurlyeq y$ can also be written as $x \wedge \bar{y} = 0$ with $\bar{y}$ the bitwise complement of $y$.

**Theorem 3.3** *Let $u$, $v$ and $w$ be masks in $\mathbb{F}_2^n$. It holds that $C^{\boxplus}_{w, u\|v} \ne 0$ if and only if $(u + w) \vee (v + w) \preccurlyeq M^{\mathsf{T}}(u + v + w)$. Furthermore, in this case*

$$C^{\boxplus}_{w,\,u\|v} = (-1)^{(u+w)^{\mathsf{T}}(v+w)} / 2^{\mathsf{wt}(M^{\mathsf{T}}(u+v+w))}.$$

*Proof* The correlation matrix of $\boxplus \colon \mathbb{F}_2^{2n} \to \mathbb{F}_2^n$ satisfies

$$C^{\boxplus}_{w, u\|v} = \frac{1}{2^n} \sum_{(x\|y, z) \in G_{\boxplus}} (-1)^{u^{\mathsf{T}} x + v^{\mathsf{T}} y + w^{\mathsf{T}} z}.$$

Using Lemma 3.2 and a substitution gives

$$C^{\boxplus}_{w, u\|v} = \frac{1}{2^n} \sum_{(x\|y, z) \in G_{\mathsf{Q}}} (-1)^{u^{\mathsf{T}}(x+z) + v^{\mathsf{T}}(y+z) + w^{\mathsf{T}}(x+y+z)}$$

$$= \frac{1}{2^n} \sum_{(x\|y, z) \in G_{\mathsf{Q}}} (-1)^{(u+w)^{\mathsf{T}} x + (v+w)^{\mathsf{T}} y + (u+v+w)^{\mathsf{T}} z}$$

$$= C^{\mathsf{Q}}_{u+v+w,\, (u+w)\|(v+w)}.$$

Since $\mathsf{Q}$ is the same as bitwise-and up to multiplication by $M$, the result follows from the formula for the correlation matrix of bitwise-and in Exercise 2.8. □

As an example, let us construct a model for one round of Speck. If the input mask is equal to $u_1 \,\|\, u_2$ and the output mask is equal to $v_1 \,\|\, v_2$, then the input mask of the modular addition is $(u_1 \ggg 8) \,\|\, (u_2 + (v_2 \ggg 3))$. The output mask is $v_1 + v_2$. Substituting this in into Theorem 3.3 gives bitvector constraints to ensure that a one-round linear approximation is effective. Combining these conditions yields the constraints for a trail to have nonzero correlation. Finally, one should add one more constraint to set the sum of the weights of the linear approximations (also given by Theorem 3.3) equal to a constant $W$. If there exists a trail with absolute correction $2^{-W}$, then the solver will find it. Otherwise, it will report that the problem is not satisfiable.

Most of the operations that occur in the abovementioned constraints are natively provided by SMT solvers that support bitvector variables. The main exceptions are multiplication by $M$ and the Hamming weight. However, expressing these in terms of bitvector operations and integer additions is simple.

### 3.3.3 Solving the model

Once a model is constructed, it can be solved with trail weight $W$ fixed to a lower bound (zero in the worst case). If the solver returns that the problem is not satisfiable, then the problem is solved again for trail weight $W + 1$. This process is repeated until a minimum-weight solution is found.

Popular SMT solvers with bitvector support include Boolector and Z3. Problems are either specified using a solver-dependent API or the widely supported LibSMT file format.

## 3.4 Historical remarks

Methods to automate the search for linear trails have been used since the early days of linear cryptanalysis, starting with Matsui's branch and bound method. Off-the-shelf MILP and SAT solvers became a popular alternative around 2010, and by now the literature on this topic is substantial.

The wide-trail strategy was introduced by Daemen in his PhD thesis, and subsequently developed in joint work with Rijmen. The block cipher Rijndael, on which the example in Section 3.2.1 is loosely based, was designed by Daemen and Rijmen in 1997, and its 128-bit version was standardized by the United States National Institute for Standards and Technology (NIST) in 2001.

Although the terminology "ARX cipher" is more recent, the first designs following this approach were published in the 1980s. The block cipher Speck that we use as an example in Section 3.3.1 appeared in 2013. An efficient algorithm to compute the correlations of linear approximations for modular addition was first described in 2003 by Wallén. The simplified formula in Theorem 3.3 is due to Schulte–Geers.

## 3.5 References

Beaulieu, Ray et al. (2013). *The SIMON and SPECK Families of Lightweight Block Ciphers*. Cryptology ePrint Archive, Report 2013/404. URL: https://eprint.iacr.org/2013/404.

Daemen, Joan and Vincent Rijmen (Dec. 2001). "The Wide Trail Design Strategy." In: *8th IMA International Conference on Cryptography and Coding*. Ed. by Bahram Honary. Vol. 2260. LNCS. Springer, Berlin, Heidelberg, pp. 222–238. DOI: 10.1007/3-540-45325-3_20.

Schulte-Geers, Ernst (2013). "On CCZ-equivalence of addition mod $2^n$." In: *Designs, Codes and Cryptography* 66, pp. 111–127.

Wallén, Johan (Feb. 2003). "Linear Approximations of Addition Modulo $2^n$." In: *FSE 2003*. Ed. by Thomas Johansson. Vol. 2887. LNCS. Springer, Berlin, Heidelberg, pp. 261–273. DOI: 10.1007/978-3-540-39887-5_20.

## 3.6 Exercises

### Exercise 3.1

This exercise explores the properties of the MixColumns step of the Rijndael-like example cipher. Recall from Section 3.2.1 that this map corresponds to the block matrix

$$M = \begin{bmatrix} 0 & I & I & I \\ I & 0 & I & I \\ I & I & 0 & I \\ I & I & I & 0 \end{bmatrix},$$

with $3 \times 3$ blocks over $\mathbb{F}_2$.

1. Show that $M$ is invertible and find its inverse.
2. What is the linear branch number of $M$?

### * Exercise 3.2

Consider the Rijndael-like example cipher from Section 3.2.1. Show that

1. Every two-round linear trail has absolute correlation at most $1/2^4$.
2. Every four-round linear trail has absolute correlation at most $1/2^{16}$.

### Exercise 3.3

This question explores some properties of the branch number $\mathcal{B}(M)$ of a matrix $M$ or linear map $x \mapsto Mx$ over $\mathbb{F}_2$ (see Definition 3.1). Suppose that $M$ is a

## 3.6 Exercises

$bm \times bn$ matrix and that the Hamming weight is defined with respect to $b$-bit blocks.

1. Prove that if $M$ is invertible, then $\mathcal{B}(M) = \mathcal{B}(M^{-1})$.
2. Using the definition of the branch number, show that the Hamming weight of every element of the vector space $C = \{u \parallel v \mid v \in \mathbb{F}_2^{bm} \text{ and } u = M^\mathsf{T} v\}$ is at least $\mathcal{B}(M)$. What is the vector space $C$ corresponding to $M^{-1}$?
3. Show that the branch number of $M$ is at most $n+1$.

In the coding theory literature, the vector space $C$ is called a *block code*. The upper bound of $n+1$ on the branch number is known as the *singleton bound*. A matrix $M$ with branch number $n+1$ is called *maximum distance separable* or MDS for short.

### * Exercise 3.4

This exercise introduces a particular construction of MDS matrices based on polynomials over finite fields. Hence, some familiarity with the theory of finite fields is required to solve this exercise. As discussed in Exercise 3.3, a $bn \times bn$ matrix over $\mathbb{F}_2$ with $b \times b$ blocks is called MDS if its branch number is equal to $\mathcal{B}(M) = n + 1$.

1. Let $\mathbb{F}_{2^b}$ be a finite field of order $2^b$ and $L\colon \mathbb{F}_{2^b}^n \to \mathbb{F}_{2^b}^n$ an arbitrary linear map. Show that there exists an invertible $\mathbb{F}_2$-linear map $\beta\colon \mathbb{F}_2^{bn} \to \mathbb{F}_{2^b}^n$ and a $bn \times bn$ matrix $M$ over $\mathbb{F}_2$ so that $x \mapsto \beta^{-1}(M^\mathsf{T} \beta(x))$ is equal to $L$. Show that the branch number of $M$ satisfies

$$\mathcal{B}(M) = \max_{x \neq 0} \text{wt}(x) + \text{wt}(L(x)),$$

with $\text{wt}(x)$ the number of nonzero coordinates of $x$ in $\mathbb{F}_{2^b}^n$.

2. Let $p_x$ be a polynomial of degree $\leq n-1$ with coefficients $x_1, \ldots, x_n$, where $x_1, \ldots, x_n$ are the coordinates of a vector $x$ in $\mathbb{F}_{2^b}^n$. Show that for all $\alpha_1, \ldots, \alpha_n$ in $\mathbb{F}_{2^b}$, the following map $L\colon \mathbb{F}_{2^b}^n \to \mathbb{F}_{2^b}^n$ is linear:

$$L\colon x \mapsto \begin{bmatrix} p_x(\alpha_1) \\ \vdots \\ p_x(\alpha_n) \end{bmatrix}.$$

3. Construct elements $\alpha_1, \ldots, \alpha_n$ of $\mathbb{F}_{2^b}$ so that the matrix $M$ constructed from the function $L$ has branch number $n+1$.

The code corresponding to $M$ (see Exercise 3.3) is called a Reed–Solomon code.

### Exercise 3.5

The S-box of the block cipher Rijndael is based on the map $x \mapsto 1/x$ in a finite field $\mathbb{F}_{2^n}$. Let $\beta\colon \mathbb{F}_{2^n} \to \mathbb{F}_2^n$ be an invertible linear map, $A$ an $n \times n$ matrix over $\mathbb{F}_2$ and $b$ a vector in $\mathbb{F}_2^n$. Define the S-box $\mathsf{S}\colon \mathbb{F}_2^n \to \mathbb{F}_2^n$ by

$$\mathsf{S}(x) = \begin{cases} A\beta^{-1}(1/\beta(x)) + b & \text{if } x \neq 0, \\ b & \text{else.} \end{cases}$$

The goal of this exercise is to show that all nontrivial linear approximations of the AES have low absolute correlation. To prove this result, you may use the following bound for *Kloosterman sums*:

$$\left| \sum_{x \in \mathbb{F}_{2^n}^\times} (-1)^{\mathrm{Tr}(x + c/x)} \right| \leq 2^{n/2+1},$$

where $\mathrm{Tr}\colon \mathbb{F}_{2^8} \to \mathbb{F}_2$ is the trace function and $c$ is a constant in $\mathbb{F}_{2^n}$. This exercise requires some familiarity with the theory of finite fields.

1. Show that it is sufficient to bound the correlations of linear approximations of the function defined by $x \mapsto \beta^{-1}(1/\beta(x))$ for $x \neq 0$ and $0 \mapsto 0$.
2. Prove that the absolute correlation of all nontrivial linear approximations of $\mathsf{S}$ is at most $2^{1-n/2}$.
3. Look up the Rijndael S-box and compute its correlation matrix. Compare with your result for $n = 8$.

### Exercise 3.6

Use branch and bound, MILP or SMT to automate the optimization of trails in the example cipher from Section 1.1. Use any one of the following tools to solve this exercise:

- Any programming language to implement the branch and bound method.
- Python with Google OR-Tools[2] to create and solve MILP models.
- Python with PySMT[3] to create and solve SMT models.

Using your model of the example cipher, solve the following tasks:

1. Verify the results from Example 2.3.
2. Find a linear trail over five rounds with correlation $\pm 2^{-5}$.

---

[2] https://developers.google.com/optimization
[3] https://github.com/pysmt/pysmt

3. Choose one of the linear trails from the previous question, and find all linear trails for the same approximation.

## Exercise 3.7

Use MILP to model the propagation of linear trails in the Rijndael-like example cipher from Section 3.2.1.

1. Based on Exercise 3.2, the maximum absolute correlation of a linear trail over four rounds is at most $2^{-16}$. Find a trail that matches this bound.
2. For the linear approximation you found above, find the linear trail(s) with the next-largest absolute correlation. Discuss the implications of these additional trails.

## Exercise 3.8

Use SMT to model the propagation of linear trails in Simon with a block size of 32 bits, as described in Exercise 2.9.

1. Verify the result you found in Exercise 2.9.
2. Are there any other linear trails for the same linear approximation that should be taken into account?

## * Exercise 3.9

Use SMT to model the propagation of linear trails in Speck with a block size of 64 bits, as described in Section 3.3.1. Find an optimal linear trail over seven rounds. Implement Matsui's first algorithm to recover one bit of information about the secret key, and test if it works as expected.

# 4
# Statistics of linear cryptanalysis

Determining the effectiveness of linear cryptanalysis is an application of statistical theory. In this chapter, we review some basic concepts from statistics and discuss how they are used to estimate the cost of linear attacks, and Matsui's second algorithm in particular.

## 4.1 Statistical inference

Parameter estimation and hypothesis testing are two closely related problems in statistical inference. Both of them are important for the analysis of linear attacks, so this section reviews the basic principles. The main result is Theorem 4.1, about hypothesis testing between two normal distributions with the same variance. It is used repeatedly throughout this book. Appendix A provides the necessary background on the normal distribution.

### 4.1.1 Estimators

Let **x** be a random variable with probability distribution $P_\theta$, with $\theta$ an unknown parameter. For example, suppose that **x** has a normal distribution with mean $\mu$ and variance $\sigma^2$, but we do not know the value of $\mu$. For brevity, this is denoted by $\mathbf{x} \sim \mathcal{N}(\mu, \sigma^2)$.

The unknown parameter $\theta$ can be estimated or "inferred" based on *samples* $x_1, \ldots, x_q$ from the distribution $P_\theta$. An *estimator* for the parameter $\theta$ of the distribution $P_\theta$ is a function $f$ that maps a sample $(x_1, \ldots, x_q)$ to an estimate of the parameter $\theta$.

## 4.1 Statistical inference

*Example* 4.1   A typical estimator for the mean $\mu = \mathbb{E}(\mathbf{x})$ of a random variable $\mathbf{x}$ is the *sample average*

$$\widehat{\mu}(x_1, \ldots, x_q) = \frac{1}{q} \sum_{i=1}^{q} x_i.$$

If the mean of the distribution of $\mathbf{x}$ is an unknown parameter, then this estimator can be used to infer its value.   ▷

To discuss the statistical properties of estimators, we consider the sample itself to be a random variable. The distribution of a random sample $(\mathbf{x}_1, \ldots, \mathbf{x}_q)$ depends on the distribution $P_\theta$ as well as on the sampling strategy. The simplest strategy is sampling with replacement: in this case, $\mathbf{x}_1, \ldots, \mathbf{x}_q$ are independent and all have distribution $P_\theta$. An estimator $f$ for $\theta$ is unbiased if

$$\mathbb{E}_{\mathbf{x}_1, \ldots, \mathbf{x}_q} f(\mathbf{x}_1, \ldots, \mathbf{x}_q) = \theta.$$

Perhaps counterintuitively, there are also cases where biased estimators are useful – in particular when $f(\mathbf{x}_1, \ldots, \mathbf{x}_q)$ is close to $\theta$ with high probability.

*Example* 4.2   The sample average defined in Example 4.1 is an unbiased estimator for the mean of a random variable $\mathbf{x}$. Indeed, let $(\mathbf{x}_1, \ldots, \mathbf{x}_q)$ be a random sample so that the marginal distribution of $\mathbf{x}_i$ matches the distribution of $\mathbf{x}$ for $i = 1, \ldots, q$. Since

$$\mathbb{E}_{\mathbf{x}_1, \ldots, \mathbf{x}_q} \widehat{\mu}(\mathbf{x}_1, \ldots, \mathbf{x}_q) = \mathbb{E}_{\mathbf{x}_1, \ldots, \mathbf{x}_q} \frac{1}{q} \sum_{i=1}^{q} \mathbf{x}_i = \frac{1}{q} \sum_{i=1}^{q} \mathbb{E}(\mathbf{x}_i) = \mathbb{E}(\mathbf{x}),$$

the sample average is an unbiased estimator for the mean of $\mathbf{x}$.   ▷

In addition to the average of an estimator, we need to consider how much the estimator deviates from its average. One measure for this is the variance

$$\mathbb{V}_{\mathbf{x}_1, \ldots, \mathbf{x}_q} f(\mathbf{x}_1, \ldots, \mathbf{x}_q) = \mathbb{E}_{\mathbf{x}_1, \ldots, \mathbf{x}_q} \left( f(\mathbf{x}_1, \ldots, \mathbf{x}_q) - \mu \right)^2,$$

where $\mu$ is the average of the estimator. If $f$ is unbiased, then $\mu = \theta$.

*Example* 4.3   Let $\sigma^2$ denote the variance of $\mathbf{x}$. If the samples $\mathbf{x}_1, \ldots, \mathbf{x}_q$ are independent random variables with the same marginal distribution as $\mathbf{x}$, then the variance of the sample average is given by

$$\mathbb{V}_{\mathbf{x}_1, \ldots, \mathbf{x}_q} \widehat{\mu}(\mathbf{x}_1, \ldots, \mathbf{x}_q) = \mathbb{V}_{\mathbf{x}_1, \ldots, \mathbf{x}_q} \frac{1}{q} \sum_{i=1}^{q} \mathbf{x}_i = \frac{1}{q^2} \sum_{i=1}^{q} \mathbb{V}(\mathbf{x}_i) = \frac{\sigma^2}{q}.$$

The third equality follows from the fact that the variance of a sum of independent random variables is the sum of the variances of the summands.   ▷

### 4.1.2 Hypothesis tests

The purpose of a *statistical hypothesis test* is to falsify a hypothesis about a probability distribution in the face of uncertainties. By convention, the hypothesis to falsify is called the null hypothesis. It is often compared to a second hypothesis, which is called the alternative hypothesis. This is the classical interpretation of statistical hypothesis testing due to Fisher. In cryptanalysis, it is more appropriate to follow the Neyman–Pearson interpretation of statistical testing as a decision (or "distinguishing") problem. In this case, the alternative hypothesis is considered to be a competing hypothesis.

A hypothesis test takes as input the value of a test statistic and outputs whether or not the null hypothesis should be rejected. The test statistic is a function of the observed data (the sample). For example, to test a hypothesis about a parameter $\theta$ of a distribution $P_\theta$, the test statistic could be an estimator for $\theta$.

Throughout this chapter, it is assumed that the test statistic $f$ is real-valued and the hypothesis test compares $t = f(x_1, \ldots, x_q)$ with a threshold value $\tau$. If $t \geq \tau$, then the test fails to falsify ("accepts") the null hypothesis. Otherwise, if $t < \tau$, the null hypothesis is rejected. Let $\mathbf{t}_{\text{null}}$ denote the value of the test statistic for a random sample obtained under the null hypothesis. Similarly, let $\mathbf{t}_{\text{alt}}$ denote the value of the test statistic for a random sample under the alternative hypothesis. For every hypothesis test, there are two important probabilities:

$$P_S = \Pr\left[\mathbf{t}_{\text{null}} \geq \tau\right],$$
$$P_F = \Pr\left[\mathbf{t}_{\text{alt}} \geq \tau\right].$$

The probability $P_S$ is called the *true-positive probability* or (in cryptanalysis) the *success probability*, whereas $P_F$ is called the *false-positive probability*. In general, there is always a trade-off between $P_S$ and $1 - P_F$, since decreasing $\tau$ will increase both $P_S$ and $P_F$ (typically by different amounts, however). This means that neither $P_S$ nor $1 - P_F$ are good measures of quality when viewed in isolation. Cryptographers sometimes use the following measure of quality, known as *advantage*:

$$|P_S - P_F|.$$

Observe that this definition is symmetric with respect to the null hypothesis and the alternative hypothesis. The concept of advantage is arbitrary. Alternative measures are discussed in Chapter 7.

In the remainder of this section, we study statistical hypothesis tests on the mean of a normal distribution. Specifically, the null hypothesis states that

## 4.1 Statistical inference

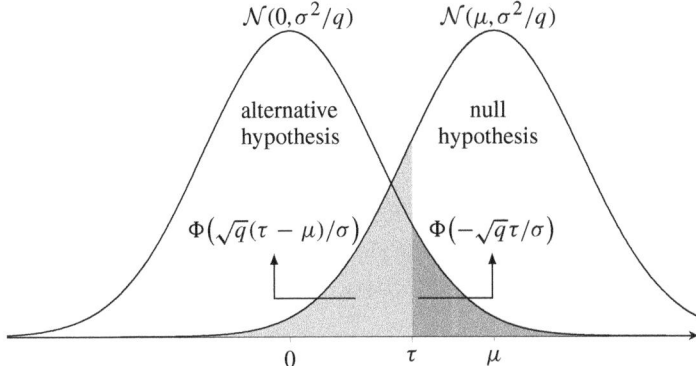

Figure 4.1 Distributions under the null and alternative hypothesis.

the distribution of the test statistic is normal with mean $\mu \neq 0$, whereas the alternative hypothesis states that it is normal with mean zero. Both hypotheses state that the standard deviation is $\sigma/\sqrt{q}$. As discussed in Example 4.3, this is the standard deviation of the sample average of $q$ samples from a distribution with standard deviation $\sigma$. The overall situation is illustrated in Figure 4.1.

The following result is important for the analysis of linear attacks. It is used throughout this book. In Theorem 4.1, $\Phi$ denotes the cumulative distribution function of the normal distribution with mean zero and variance one.

**Theorem 4.1** *For a hypothesis test that compares a sample to a threshold in order to distinguish two normal distributions with the same variance $\sigma^2/q$ and means $\mu \neq 0$ (null hypothesis) and zero (alternative hypothesis), the success probability $P_S$ and false-positive probability $P_F \leq P_S$ satisfy the relation*

$$q = \left( \frac{\Phi^{-1}(P_S) - \Phi^{-1}(P_F)}{\mu/\sigma} \right)^2.$$

*Proof* Assume that $\mu > 0$. A similar reasoning holds for $\mu < 0$. As illustrated in Figure 4.1, the test accepts the null hypothesis when the test statistic (equal to the sample value itself) exceeds $\tau$. As above, let $\mathbf{t}_{\text{null}}$ denote the value of the test statistic for a random sample obtained under the null hypothesis, and $\mathbf{t}_{\text{alt}}$ its value for a random sample from the alternative hypothesis.

Under the null hypothesis, the test statistic has a normal distribution with mean $\mu$ and variance $\sigma^2/q$. The success probability is the probability that $\mathbf{t}_{\text{null}}$ is larger than $\tau$ when the null hypothesis is true:

$$P_S = \Pr\left[\mathbf{t}_{\text{null}} \geq \tau\right] = 1 - \Phi\big((\tau - \mu)/(\sigma/\sqrt{q})\big) = \Phi\big(\sqrt{q}\,(\mu - \tau)/\sigma\big).$$

Hence, $\sqrt{q}\tau/\sigma = \sqrt{q}\mu/\sigma - \Phi^{-1}(P_S)$. The false-positive probability is the probability that $t_{alt}$ is larger than $\tau$ when the alternative hypothesis is true:
$$P_F = \Pr[t_{alt} \geq \tau] = 1 - \Phi(\tau/(\sigma/\sqrt{q})) = \Phi(-\sqrt{q}\,\tau/\sigma).$$
Equivalently, $\sqrt{q}\tau/\sigma = -\Phi^{-1}(P_F)$. From the two expressions for $\sqrt{q}\tau/\sigma$,
$$\Phi^{-1}(P_S) = \sqrt{q}\,\mu/\sigma + \Phi^{-1}(P_F).$$
Rearranging the terms and squaring yields the result, provided that $\Phi^{-1}(P_S) \geq \Phi^{-1}(P_F)$. Since $\Phi$ is strictly increasing, the latter condition is equivalent to $P_S \geq P_F$. $\square$

## 4.2 Key-recovery using statistical hypothesis testing

Recall from Section 1.4.2 that Matsui's Algorithm 2 computes the empirical correlation of a linear approximation over the inner part of a cipher for all possible values of the relevant key bits in the outer part of the cipher. This approach assumes that the empirical correlation is larger (in absolute value) for the correct key than for incorrect keys.

From the point of view of statistical hypothesis testing, for every empirical correlation computed by Matsui's Algorithm 2, we need to decide whether or not the corresponding key is correct. If there are $K$ possible keys, then filtering out the most promising candidates requires $K$ hypothesis tests. Suppose all of these tests have success probability $P_S$ and false-positive probability $P_F$. By definition, the probability that the correct key is recovered equals $P_S$. Furthermore, the average number of candidate keys is $P_S + P_F(K-1) \approx P_F K$.

The following two sections analyze the number of known plaintexts required by Matsui's Algorithm 2 in order to recover the correct partial key with probability $P_S$ as one of approximately $P_F K$ candidate keys. Modulo the approximations that are discussed below, this is the same as the data-complexity of a distinguisher with success probability $P_S$ and false-positive probability $P_F$ based on a linear approximation of the inner part of the cipher. Section 4.2.1 considers the case where this linear approximation has known (key-independent) correlation. The case with unknown correlations is discussed in Section 4.2.2.

### 4.2.1 Known correlation

Typical for cryptanalysis is that one does not completely know the distributions that the statistical test is supposed to tell apart. To make it feasible to analyze

## 4.2 Key-recovery using statistical hypothesis testing

the cost of linear distinguishers, we propose a *model*, i.e. we presume certain distributions and make some additional approximations. This chapter starts with a rather coarse model, called the "simple model." Refinements of the simple model are discussed in Chapter 7.

In the simple model, the null hypothesis states that the correlation of the linear approximation is equal to $c \neq 0$. The alternative hypothesis states that the correlation is *exactly zero*. The latter hypothesis has its origins in the assumption that incorrect keys lead to empirical correlations that are closer to zero, which is also known as the *wrong key randomization hypothesis*. Intuitively, a wrong guess "randomizes" the test statistic, which should result in a correlation (close to) zero.

In Chapter 7, it will be shown that even linear approximations over a uniform random permutation or function rarely have correlation exactly zero. Consequently, the simple model is not necessarily an accurate representation of reality, and the results obtained in this model should be understood as approximations.

The simple model makes the following additional technical assumptions:

**Small correlation:** The squared correlation is negligible compared to one.

**Large data:** The amount of data $q$ is large enough so that, e.g., the normal approximation to the binomial distribution is accurate.

**Sampling model:** Plaintexts are sampled uniformly at random *with replacement*. This implies, in particular, that samples are independent.

These assumptions are usually realistic, and they can be avoided without too much difficulties if necessary (see for instance Exercise 4.1).

In the simple model, the data-complexity of a linear distinguisher is essentially given by Theorem 4.1. For a uniform random input $\mathbf{x}$ and corresponding output $\mathbf{y} = F(\mathbf{x})$, define the random variable

$$\mathbf{z} = (-1)^{u^T \mathbf{x} + v^T \mathbf{y}}.$$

By definition, $\mathbb{E}(\mathbf{z})$ is the correlation of the linear approximation $(u, v)$ of $F$. Under the null hypothesis, the correlation equals $c$. Under the alternative hypothesis, it equals zero. Furthermore, $\mathbb{E}(\mathbf{z}^2) = 1$ for either hypothesis. Hence, under the null hypothesis, $\mathbb{V}(\mathbf{z}) = 1 - c^2 \approx 1$ and under the alternative $\mathbb{V}(\mathbf{z}) = 1$.

Collecting $q$ known plaintexts and their corresponding ciphertexts, or rather the values $\mathbf{z}_1 = (-1)^{u^T \mathbf{x}_1 + v^T \mathbf{y}_1}, \ldots, \mathbf{z}_q = (-1)^{u^T \mathbf{x}_q + v^T \mathbf{y}_q}$, corresponds to sampling $\mathbf{z}$. The statistical tests we perform are based on the sample average or *empirical correlation*

$$\widehat{\mathbf{c}} = \frac{1}{q} \sum_{i=1}^{q} \mathbf{z}_i.$$

As discussed in Section 4.1.1, the sample average is an estimator of the mean. In other words, the empirical correlation is a statistical estimator of the correlation.

Since the inputs are sampled with replacement, the random samples $\mathbf{z}_1, \ldots, \mathbf{z}_q$ are independent under either hypothesis. If $q$ is large, then the central limit theorem (Theorem A.1 in Appendix A) implies that $\widehat{\mathbf{c}}$ approximately follows a normal distribution. By Examples 4.2 and 4.3, the mean of $\widehat{\mathbf{c}}$ is equal to the population mean, i.e. 0 or $c$, and the variance of $\widehat{\mathbf{c}}$ is equal to the variance of the population divided by the sample size $q$, i.e. $1/q$. Hence, Theorem 4.1 implies the following result.

**Corollary 4.2** *The data complexity of a linear distinguisher in the simple model, using a linear approximation with known correlation $c$, is*

$$q = \left( \frac{\Phi^{-1}(P_S) - \Phi^{-1}(P_F)}{c} \right)^2,$$

*where $P_S$ is the success probability and $P_F \leq P_S$ the false-positive probability.*

*Proof* The result follows by setting $\mu = c$ and $\sigma = 1$ in Theorem 4.1. □

### 4.2.2 Unknown correlation

For most of the attacks described in the literature, the correlation of the linear approximation depends on the (unknown) key. Furthermore, it is often difficult to compute the correlation of linear approximations exactly even when the key is known, because there can be many linear trails that contribute to the correlation.

When the correlation is unknown, the main difficulty in applying the test from Section 4.2.1 is that the sign of the correlation is not known. There are two possible solutions to this problem. One strategy is to examine both the positive and negative cases, performing two hypothesis tests for every possible key. This increases the false-positive probability, approximately doubling the number of candidate keys. Another strategy is to work with a test statistic that does not depend on the sign of $c$. For example, one can develop a test based on $|\widehat{\mathbf{c}}|$ or equivalently $\widehat{\mathbf{c}}^2$. The null hypothesis is accepted if the absolute or squared correlation exceeds a threshold $\tau$. This test works as long as the absolute correlation is significantly larger under the null hypothesis than under the alternative hypothesis.

## 4.2 Key-recovery using statistical hypothesis testing

The following theorem gives the success probability when $|c|$ is known, but the sign of $c$ is not known. For key-alternating ciphers, this corresponds to the case of linear approximations that are dominated by the correlation of a single linear trail. The general case, when $|c|$ is also key-dependent, is discussed afterwards. Theorem 4.3 makes the same assumptions as the simple model, except that the test is based on the absolute value or square of the empirical correlation.

**Theorem 4.3** *The success probability of a linear distinguisher in the simple model based on the absolute value or square of the empirical correlation, and using a linear approximation with known absolute correlation $|c|$, is*

$$P_S = \Phi\bigl(\Phi^{-1}(P_F/2) + |c|\sqrt{q}\bigr) + \Phi\bigl(\Phi^{-1}(P_F/2) - |c|\sqrt{q}\bigr),$$

*where $q$ is the data-complexity and $P_F$ the false-positive probability.*

*Proof* The empirical correlation $\widehat{c}_{\text{null}}$ under the null hypothesis has a normal distribution with mean $c$ and variance $1/q$. Similarly, the empirical correlation $\widehat{c}_{\text{alt}}$ under the alternative hypothesis has mean zero and variance $1/q$.

Since $\widehat{c}^2 \geq \tau$ is equivalent to $|\widehat{c}| \geq \sqrt{\tau}$ for $\tau \geq 0$, it does not matter whether we use the absolute value of the correlation or its square. Hence, for $\tau \geq 0$, the false-positive probability $P_F$ is equal to

$$P_F = \Pr\bigl[|\widehat{c}_{\text{alt}}| \geq \tau\bigr] = 2\Phi\bigl(-\sqrt{q}\tau\bigr).$$

Solving for $\tau$ yields $\sqrt{q}\tau = -\Phi^{-1}(P_F/2)$. The success probability $P_S$ equals

$$\begin{aligned} P_S &= \Pr\bigl[|\widehat{c}_{\text{null}}| \geq \tau\bigr] \\ &= \Pr\bigl[\widehat{c}_{\text{null}} \geq \tau\bigr] + \Pr\bigl[\widehat{c}_{\text{null}} \leq -\tau\bigr] \\ &= \Phi\bigl(-\sqrt{q}(\tau - c)\bigr) + \Phi\bigl(-\sqrt{q}(\tau + c)\bigr). \end{aligned}$$

Since the above expression is an even function of $c$, it can be rewritten as

$$P_S = \Phi\bigl(-\sqrt{q}\tau + |c|\sqrt{q}\bigr) + \Phi\bigl(-\sqrt{q}\tau - |c|\sqrt{q}\bigr).$$

Plugging in $\tau\sqrt{q} = -\Phi^{-1}(P_F/2)$ yields the result. □

If the false-positive probability $P_F$ is small enough, then the expression for the success probability in Theorem 4.3 is well approximated by $P_S \approx \Phi(\Phi^{-1}(P_F/2) + |c|\sqrt{q})$. Hence, in this case, the data-complexity is given by

$$q \approx \left(\frac{\Phi^{-1}(P_S) - \Phi^{-1}(P_F/2)}{c}\right)^2,$$

assuming that $P_S \geq P_F/2$.

Theorem 4.3 can still be used even if the absolute value of the correlation depends on the key. In Chapter 7, it is shown that the test is optimal when only the sign is key-dependent, but not in the general case. To analyze the success probability when $|c|$ depends on the key, the expression for $P_S$ in Theorem 4.3 should be averaged with respect to the key. The following example illustrates this.

*Example* 4.4 (Revisiting Example 2.3)  This example determines the success probability of a distinguisher based on the linear approximation (000000001, 000010000) for three rounds of the example cipher from Section 1.1. In Example 2.3, the following expression for the correlation was derived:

$$c = (-1)^{\kappa_1}/8\left(1 + (-1)^{\kappa_2}/2\right)\left(1 + (-1)^{\kappa_3}/2\right),$$

or for the squared correlation,

$$c^2 = \left(\left(1 + (-1)^{\kappa_2}/2\right)\left(1 + (-1)^{\kappa_3}/2\right)\right)^2/64.$$

Hence, $c^2 = 1/2^{10}$ for 25% of the keys, $c^2 = 9/2^{10}$ for 50% of the keys and $c^2 = 81/2^{10}$ for 25% of the keys. Figure 4.2 shows the success probability $P_S$ as a function of $q$, for constant false-positive probability. The success probability is the average of the formula from Theorem 4.3 with respect to the key. ▷

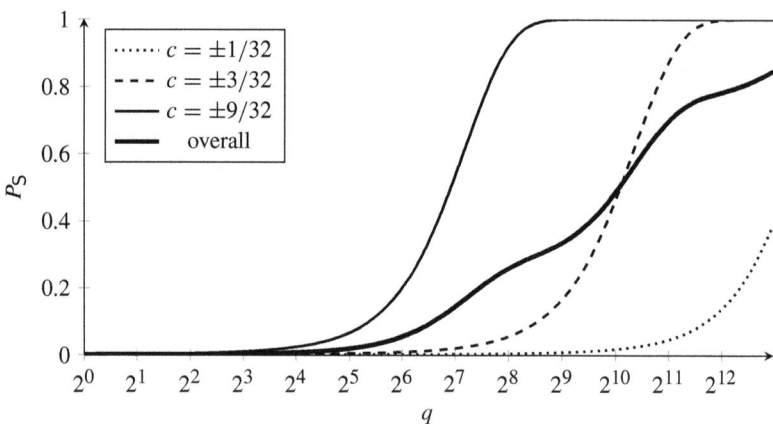

Figure 4.2 The success probability $P_S$ as a function of $q$ for the linear distinguisher from Example 4.4, with $P_F = 0.002$. The overall success probability curve is a weighted sum of the curves for the three subsets of keys, i.e., overall$(q) = 1/4\,\text{case}_1(q) + 1/2\,\text{case}_2(q) + 1/4\,\text{case}_3(q)$.

## 4.3 Sampling strategies

In the previous sections, it was assumed that plaintexts are sampled uniformly at random with replacement. In this case the samples are independent and $|\{1 \le i \le q \mid u^\mathsf{T}\mathbf{x}_i = v^\mathsf{T}\mathbf{y}_i\}| = \sum_{i=1}^{q}(\mathbf{z}_i + 1)/2$ follows a binomial distribution, which we approximated by a normal distribution based on the central limit theorem.

If plaintexts are sampled without replacement, then the random variables $\mathbf{z}_i$ are not independent: with each draw of the value $+1$, the probability that the next draw results in $+1$ is decreased, and vice versa. It can be shown that in this case, $|\{1 \le i \le q \mid u^\mathsf{T}\mathbf{x}_i = v^\mathsf{T}\mathbf{y}_i\}|$ has a hypergeometric distribution, which one can also approximate by a normal distribution. Compared to the normal approximation of the binomial distribution, the mean is the same but the variance becomes smaller: it is multiplied by the factor

$$\frac{2^n - q}{2^n - 1} \approx 1 - \frac{q}{2^n},$$

which becomes small when $q$ approaches $2^n$.

Sampling with replacement leads to simpler formulas, but sampling without replacement leads to formulas that predict a lower data-complexity. In practice, when the amount of samples $q$ is large, it could be difficult to ensure that plaintexts are unique, since the attacker might need to keep track of which values were already encountered before. This practical problem does not occur if the mode of operation of the block cipher ensures that there are no repetitions. For example, this is the case in counter mode and in other modes derived from it, such as Galois Counter Mode (GCM).

## 4.4 Key-recovery using key ranking

There is an alternative approach to key-recovery that is often used in practice. Instead of performing a hypothesis test on the empirical correlation for every key, output a sorted list of candidate keys in order of decreasing confidence (decreasing absolute empirical correlation). This approach is called *key ranking*.

Key ranking is not fundamentally different, since in practice only the highest ranked portion of the table of keys is retained. As before, the full attack typically involves a final step where the remaining unknown bits are guessed and verified.

An important advantage of key ranking is that it is more robust against inaccuracies in the model. Indeed, for ciphers constructed from building blocks with low nonlinearity, such as ARX ciphers, the wrong key randomization hypothesis that is used in the simple model deviates from reality.

Accurately analyzing the data-complexity of key ranking is more difficult than for hypothesis testing. However, if additional assumptions are made, then the analysis becomes similar. More specifically, key-ranking can be analyzed using *order statistics*. Let $|\hat{c}_1|, |\hat{c}_2|, \ldots, |\hat{c}_K|$ be the absolute values of the empirical correlations for the $K$ keys. The order statistics are the random variables $\mathbf{s}_1, \mathbf{s}_2, \ldots, \mathbf{s}_K$ obtained by sorting $|\hat{c}_1|, |\hat{c}_2|, \ldots, |\hat{c}_K|$ such that $\mathbf{s}_1 \leq \mathbf{s}_2 \leq \cdots \leq \mathbf{s}_K$ with probability one. In particular, $\mathbf{s}_i$ is called the $i$th order statistic.

Suppose that empirical correlations for different incorrect keys are independent and have identical distributions. If the number of keys $K$ is large enough, then the order statistics look a lot like the quantile function (inverse of the cumulative distribution function) of the absolute value $|\hat{c}_i|$ of the empirical correlations for incorrect keys. In the simple model, this quantile function is $p \mapsto \Phi^{-1}((p-1)/2)/\sqrt{q}$. In particular, the $i$th order statistic satisfies

$$\mathbf{s}_i \approx \frac{1}{\sqrt{q}} \Phi^{-1}\left(\frac{i-K}{2K}\right).$$

In the above, the "$\approx$" sign signifies that $\mathbf{s}_i$ is close to the right-hand side with high probability.

The fraction of keys that are retained as candidate keys is usually written as $2^{-a}$, where $a$ is called the *key-recovery advantage* (not to be confused with advantage in the sense of hypothesis testing!). Key ranking is successful if the absolute value of the empirical correlation is larger than the $\lfloor K(1-2^{-a})\rceil$th order statistic, i.e.,

$$\mathbf{s}_{\lfloor K(1-2^{-a})\rceil} \approx \frac{1}{\sqrt{q}} \Phi^{-1}\left(2^{-a-1}\right).$$

However, this is exactly the same as the threshold value for hypothesis testing with $P_\mathsf{F} = 2^{-a}$. Hence, under the same assumptions as in Section 4.2.2,

$$q \approx \left(\frac{\Phi^{-1}(P_\mathsf{S}) - \Phi^{-1}(2^{-a-1})}{c}\right)^2.$$

Recall from Section 4.2 that $P_\mathsf{F} K$ was also a good approximation for the average number of remaining keys in the hypothesis testing approach.

## 4.5 Historical remarks

In his paper applying linear cryptanalysis to the block cipher DES, Matsui provided an estimate of the data-complexity and success probability of his attack. Selçuk analyzed the key-ranking procedure in more detail, relying on the assumptions of the simple model from Section 4.2. The data-complexity for sampling without replacement was analyzed by the authors in (Ashur, Beyne, and Rijmen 2020) and, independently, by Blondeau and Nyberg under the name "distinct known plaintext model."

There is a large body of work on the statistics of linear cryptanalysis, dealing with refinements of the simple model and optimal methods for hypothesis testing. These topics are discussed in Chapter 7.

## 4.6 References

Ashur, Tomer, Tim Beyne, and Vincent Rijmen (Apr. 2020). "Revisiting the Wrong-Key-Randomization Hypothesis." In: *Journal of Cryptology* 33.2, pp. 567–594. DOI: 10.1007/s00145-020-09343-2.

Blondeau, Céline and Kaisa Nyberg (2017). "Joint Data and Key Distribution of Simple, Multiple, and Multidimensional Linear Cryptanalysis Test Statistic and Its Impact to Data Complexity." In: *Designs, Codes and Cryptography* 82, pp. 319–349.

Matsui, Mitsuru (May 1994a). "Linear Cryptanalysis Method for DES Cipher." In: *EUROCRYPT'93*. Ed. by Tor Helleseth. Vol. 765. LNCS. Springer, Berlin, Heidelberg, pp. 386–397. DOI: 10.1007/3-540-48285-7_33.

Selçuk, Ali Aydin (Jan. 2008). "On Probability of Success in Linear and Differential Cryptanalysis." In: *Journal of Cryptology* 21.1, pp. 131–147. DOI: 10.1007/s00145-007-9013-7.

## 4.7 Exercises

### Exercise 4.1

Modify Corollary 4.2 so it remains valid without the "small correlation" assumption from the simple model.

### Exercise 4.2

Modify Theorem 4.1, Corollary 4.2 and Theorem 4.3 for the case that plaintexts are sampled without replacement.

# 5
# Key-recovery techniques

In Chapter 1, it was explained how linear approximations can be used to set up key-recovery attacks using Matsui's Algorithm 1 or 2. This chapter takes a closer look at Algorithm 2 and its improvements. The most important improvement, and the main topic of this chapter, is the "fast Fourier transformation method."

## 5.1 Key-recovery using Algorithm 2

Recall from Chapter 1 that a key-recovery attack using Matsui's Algorithm 2 subdivides the cipher into an inner and an outer part, as shown in Figure 1.4. If the linear approximation over the inner part has sparse masks, then it can often be evaluated given a small part of the plaintext, ciphertext and round keys. This leads to a partial encryption and decryption process as illustrated in Figure 5.1, where $F_l$ and $B_k$ have a smaller domain and codomain than the inner part E.

Before introducing improvements to the naive approach of partially encrypting or decrypting every plaintext-ciphertext pair for every possible key, it is worth systematizing key-recovery attacks by introducing some terminology for the main steps in the process:

**Modeling:** The first step is finding a suitable linear approximation and determining its correlation (up to some modeling error), including how it depends on the key. This was covered in Chapters 2 and 3.

**Distillation:** As illustrated in Figure 5.1, not the entire plaintext and ciphertext are necessary to evaluate the linear approximation over the inner part of the cipher. Distillation extracts the relevant information from the plaintext-ciphertext pairs. Although this step is trivial in the naive approach, it is at the core of the improvements discussed in Sections 5.2 and 5.3.

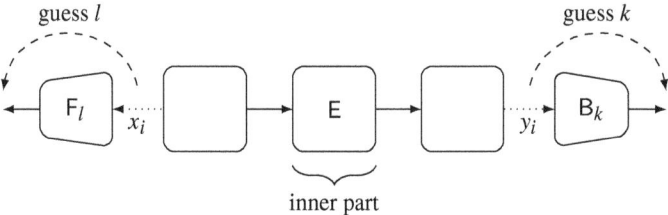

Figure 5.1 Partial encryption and decryption of the outer part of a cipher.

**Analysis:** In this step, the data are analyzed to produce a list of the most likely (partial) keys. Typically, this involves computing a test statistic for every possible key. The list of test statistics is then sorted, or compared with a predetermined threshold value. The statistical aspects of this process were discussed in Chapter 4.

**Search:** In the final step of a key-recovery attack, the full key is determined by exhaustive search over the list of remaining candidates.

## 5.2 Matsui's approach

Matsui's presentation of Algorithm 2 included an optimization that relies on a more careful distillation phase. Assume that $q$ plaintext-ciphertext pairs are available, and let us denote the $i$th truncated sample by $(x_i, y_i)$ in $\mathbb{F}_2^s \times \mathbb{F}_2^t$. Here, "truncated" refers to the fact that $x_i$ and $y_i$ only include the bits that are necessary to evaluate the linear approximation on the inner part of the cipher.

Following the same notation as in Figure 5.1, there are families of functions $\mathsf{F}_l \colon \mathbb{F}_2^s \to \mathbb{F}_2$ and $\mathsf{B}_k \colon \mathbb{F}_2^t \to \mathbb{F}_2$ such that the estimated correlation of the linear approximation over the inner part of the cipher is equal to

$$\widehat{c}_{k,l} = \frac{1}{q} \sum_{i=1}^{q} (-1)^{\mathsf{F}_l(x_i) + \mathsf{B}_k(y_i)} = \frac{1}{q} \sum_{i=1}^{q} a_l(x_i)\, b_k(y_i), \qquad (5.1)$$

where $a_l(x_i) = (-1)^{\mathsf{F}_l(x_i)}$ and $b_k(x_i) = (-1)^{\mathsf{B}_l(x_i)}$. The naive approach evaluates (5.1) for each of the $K$ possible values of $k$ and for each of the $L$ possible values of $l$, leading to a total of time-complexity of $qKL$ partial encryption and decryption operations. The optimization alluded to above improves over this when $q$ is larger than $2^{\min\{s,t\}}$.

### 5.2.1 Unidirectional

First consider the case that the outer part only consists of one or more rounds at the end of the cipher. That is, there is only $L = 1$ possible value for the partial key $l$. To emphasize this, we drop the $l$ subscripts in (5.1):

$$\widehat{c}_k = \frac{1}{q} \sum_{i=1}^{q} a(x_i) \, b_k(y_i),$$

where $a(x_i) = (-1)^{u^T x_i}$ for some mask $u$. The right-hand side above can be rewritten by grouping the terms where $y_i$ takes the same value together. This gives the equation

$$\widehat{c}_k = \sum_{y \in \mathbb{F}_2^t} b_k(y) \frac{1}{q} \sum_{i=1}^{q} a(x_i) \, \delta^y(y_i).$$

The sum above can be interpreted as a matrix-vector product. Indeed, define a $K \times 2^t$ matrix $B$ with coordinates indexed by partial keys $k$ and values $y$ and a vector $w$ by

$$B_{k,y} = b_k(y),$$

$$w_y = \frac{1}{q} \sum_{i=1}^{q} a(x_i) \, \delta^y(y_i).$$

With these definitions, the vector $\widehat{c}$ with coordinates $\widehat{c}_k$ is equal to $\widehat{c} = Bw$. This leads to the following distillation and analysis phases:

**Distillation:** Compute the vector $w$. This requires $q$ evaluations of $a$, memory accesses and additions. Storing the vector $w$ requires storing $2^t$ numbers.

**Analysis:** Compute the matrix-vector product $\widehat{c} = Bw$. The matrix-vector product can be computed without storing the matrix $B$. The computational cost is dominated by $2^t K$ partial decryptions (evaluations of $b_k$).

The overall asymptotic time-complexity is $\mathcal{O}(2^t K + q)$, compared to $\mathcal{O}(qK)$ for the naive method.

*Example* 5.1 This example uses the three-round linear approximation from Examples 1.3 and 2.3 to set up a key-recovery attack on four rounds. Figure 5.2 shows the three-round linear approximation (nonzero masks as thick lines) and the bits that are involved in partial decryption (thick lines). Based on the figure, three ciphertext bits must be known and three bits of the last round key must be

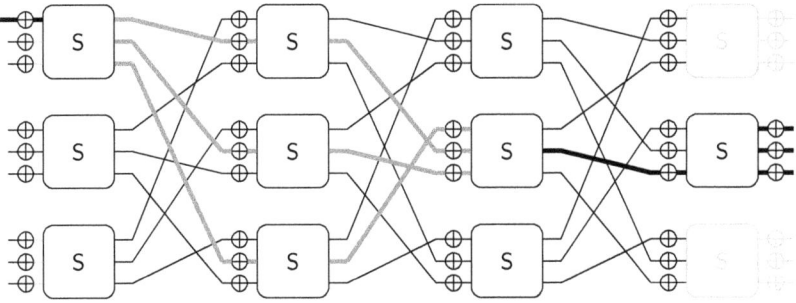

Figure 5.2 Key-recovery attack on four rounds of the example cipher.

guessed. Hence, $K = 8$ and $t = 3$. Since the correlation of the approximation is close to $1/8$, set $q = 64$.

Using 64 random samples encrypted under the all-zero key, the analysis phase involves the following matrix-vector product (indices in $\mathbb{F}_2^3$, ordered lexicographically):

$$\frac{1}{64}\underbrace{\begin{bmatrix} -24 \\ -18 \\ -12 \\ -14 \\ 24 \\ 18 \\ 12 \\ 14 \end{bmatrix}}_{\widehat{c}} = \underbrace{\begin{bmatrix} -1 & -1 & 1 & -1 & 1 & 1 & -1 & 1 \\ -1 & -1 & -1 & 1 & 1 & 1 & 1 & -1 \\ 1 & -1 & -1 & -1 & -1 & 1 & 1 & 1 \\ -1 & 1 & -1 & -1 & 1 & -1 & 1 & 1 \\ 1 & 1 & -1 & 1 & -1 & -1 & 1 & -1 \\ 1 & 1 & 1 & -1 & -1 & -1 & -1 & 1 \\ -1 & 1 & 1 & 1 & 1 & -1 & -1 & -1 \\ 1 & -1 & 1 & 1 & -1 & 1 & -1 & -1 \end{bmatrix}}_{B} \times \frac{1}{64}\underbrace{\begin{bmatrix} 5 \\ 6 \\ 1 \\ 1 \\ -6 \\ -4 \\ -4 \\ -7 \end{bmatrix}}_{w}.$$

Based on Example 2.3, the correlation for the all-zero key is $-9/32$. Hence, the empirical correlation $-24/64 = -9/32 - 3/32$ for the correct key is a plausible result. If the sign of the correlation is not known, the most likely candidate keys are 000 and 100.

Note that the wrong key randomization hypothesis mentioned in Chapter 4 does not hold for this example: the correlations for most of the incorrect keys are not close to zero, but about $\pm 1/2$ times the correlation for the correct key. This is because partial decryption with an incorrect key essentially adds one more (key-dependent) S-box to the cipher, and most effective linear approximations over this S-box have correlation $\pm 1/2$. The large correlation for incorrect key 100 is more surprising; explaining it is Exercise 5.1.  ▷

### 5.2.2 Bidirectional

The optimization introduced in Section 5.2.1 generalizes to the case where the outer part consists of one or more rounds at the beginning *and* at the end of the cipher. In this case, rewrite (5.1) as follows:

$$\widehat{c}_{k,l} = \frac{1}{q}\sum_{i=1}^{q} a_l(x_i)\, b_k(y_i) = \sum_{(x,y)\in \mathbb{F}_2^s \times \mathbb{F}_2^t} a_l(x)\, b_k(y) \frac{1}{q}\sum_{i=1}^{q} \delta^x(x_i)\delta^y(y_i).$$

Define a $2^s \times L$ matrix $A$, a $K \times 2^t$ matrix $B$ and a $2^t \times 2^s$ matrix $W$ by

$$A_{x,l} = a_l(x),$$
$$B_{k,y} = b_k(x),$$
$$W_{y,x} = \frac{1}{q}\sum_{i=1}^{q} \delta^x(x_i)\delta^y(y_i).$$

Using these definitions, the $K \times L$ matrix $\widehat{c}$ with coordinates $\widehat{c}_{k,l}$ is equal to the matrix product $BWA$. Hence, the distillation and analysis phases should be updated as follows:

**Distillation:** Compute the matrix $W$. This requires $q$ evaluations of $a$ and $b$, memory accesses and additions. Storing the matrix $W$ requires storing $2^{s+t}$ numbers.

**Analysis:** Compute the matrix product $\widehat{c} = BWA$. The matrix product can be computed either as $(BW)A$ or as $B(WA)$. The computational cost is thus

$$\min\left\{2^{s+t}LT_a + 2^t KLT_b,\, 2^{s+t}KT_b + 2^s KLT_a\right\},$$

where $T_a$ and $T_b$ are the cost of partial encryption and decryption, respectively. This requires storing at most $KL + \max\{2^t L, 2^s K\}$ numbers.

Typically $K \geq 2^t$ and $L \geq 2^s$, so that the asymptotic computational complexity is $\mathcal{O}(KL\, 2^{\min\{s,t\}} + q)$. This improves over the naive approach whenever $q \geq 2^{\min\{s,t\}}$.

In some cases, the key schedule imposes relations between the keys $k$ and $l$. These can be used to prune the matrix $\widehat{c}$. Taking this into account during the computation of $\widehat{c}$ can yield additional speed ups.

## 5.3 Fast Fourier transformation method

The most expensive part of the optimized approach to key-recovery from Section 5.2 is the calculation of the matrix-vector product $Bw$ in the

## 5.3 Fast Fourier transformation method

unidirectional case and the matrix product $BWA$ in the bidirectional case. It turns out that the matrices $A$ and $B$ often have a special structure that makes it possible to speed up these operations.

### 5.3.1 Circulant structure

If round keys are added to the state at the beginning and end of a cipher, then the functions $\mathsf{F}_l$ and $\mathsf{B}_k$ are of the form $\mathsf{F}_{l_1 \| l_2}(x) = \mathsf{F}'_{l_2}(x+l_1)$ and $\mathsf{B}_{k_1 \| k_2}(y) = \mathsf{B}'_{k_2}(y+k_1)$ for some functions $\mathsf{F}'_{l_2}$ and $\mathsf{B}'_{k_2}$ indexed by keys $l_2$ and $k_2$. Hence, there also exist $a'_{l_2}$ and $b'_{k_2}$ such that

$$a_{l_1 \| l_2}(x) = a'_{l_2}(x + l_1),$$
$$b_{k_1 \| k_2}(y) = b'_{k_2}(y + k_1).$$

This implies that the matrices $A$ and $B$ have a peculiar structure. More precisely, for all $l_2$, let $A^{l_2}$ be the square submatrix of $A$ with coordinates $A^{l_2}_{x,l_1} = a'_{l_2}(x + l_1)$. For all $k_2$, one can define $B^{k_2}$ similarly as the square submatrix of $B$ with $B^{k_2}_{k_1,y} = b'_{k_2}(y + k_1)$. The matrices $A^{l_2}$ and $B^{k_2}$ are called *circulant matrices*.

**Definition 5.1** (Circulant matrix)   A $2^m \times 2^m$ matrix $M$ with coordinates indexed by elements of $\mathbb{F}_2^m$ such that $M_{x,y} = M_{0,x+y}$ for all $x$ and $y$ in $\mathbb{F}_2^m$ is called circulant.

There exists a more general definition of circulant matrices that replaces $\mathbb{F}_2^m$ by an arbitrary finite group. You might already be familiar with circulant matrices indexed by the cyclic group $\mathbb{Z}/N\mathbb{Z}$. In this section, we limit the discussion to Definition 5.1.

*Example* 5.2 (Circulant matrix)   The matrix $B$ in Example 5.1 is circulant. For example, the first two rows of the matrix are

$$\begin{bmatrix} -1 & -1 & 1 & -1 & 1 & 1 & -1 & 1 \\ -1 & -1 & -1 & 1 & 1 & 1 & 1 & -1 \end{bmatrix}.$$

Compared to the first row, the entries at consecutive even and odd positions (counting from zero) are swapped. This corresponds to adding 001 to the column indices as elements of $\mathbb{F}_2^3$.   ▷

It turns out that a matrix-vector product with a $2^m \times 2^m$ circulant matrix can be computed in $\mathcal{O}(m 2^m)$ arithmetic operations. This assumes that one row of the matrix is stored in memory. As a result, the analysis phase can be sped up:

**Unidirectional:** To compute $Bw$, it suffices to compute the matrix-vector products $B^{k_2} w$ for all values of $k_2$. Each matrix-vector product requires

$2^t$ partial decryptions and $\mathcal{O}(t2^t)$ arithmetic operations. Hence, the time-complexity is dominated by $2^t K_2 = K_1 K_2$ partial decryption and $\mathcal{O}(t2^t K_2)$ arithmetic operations. This should be compared to the complexity $2^t K_1 K_2$ of the method from Section 5.2.

**Bidirectional:** The product $B^{k_2} W A^{l_2}$ can be computed as $(B^{k_2} W) A^{l_2}$ or $B^{k_2}(W A^{l_2})$. Without loss of generality, consider the former case. The product $B^{k_2} W$ is computed by multiplying the circulant matrix $B^{k_2}$ with the $2^s$ columns of $W$. For the product $(B^{k_2} W) A^{l_2}$, the $2^t$ rows of $B^{k_2} W$ are multiplied with the circulant matrix $A^{l_2}$. Hence, the overall cost is dominated by $K_1 K_2 L_1 L_2$ partial encryptions *or* decryptions and $\mathcal{O}(\min\{s,t\} K_1 K_2 L_1 L_2)$ arithmetic operations.

It will be clear from Section 5.3.2 that many of the arithmetic operations can be amortized away when $K_2$ and/or $L_2$ are large, although the overall complexity does not change.

If $K = K_1 K_2$ and $L = L_1 L_2$, then the overall time-complexity is $\mathcal{O}(KL)$ partial encryptions and decryptions and $\mathcal{O}(\min\{s,t\} KL)$ arithmetic operations. Compare this with the complexity $\mathcal{O}(2^{\min\{s,t\}} KL)$ of the method from Section 5.2.

When the above method is used, it is more difficult to prune the matrix $\widehat{c}$ to account for potential relations between $K_1$ and $L_1$ introduced by the keys-schedule. However, linear relations can still be taken into account by a suitable modification of the matrix-multiplication algorithm.

### 5.3.2 Multiplication with circulant matrices

Multiplication with a circulant matrix can be done efficiently using the fast Fourier transformation algorithm. This is because, as shown by Theorem 5.3 below, the Fourier transformation diagonalizes circulant matrices. The Fourier transformation (for the additive group $\mathbb{F}_2^m$) is defined as follows.

**Definition 5.2** (Fourier transformation) The Fourier transformation $\mathcal{F}_m$ is a linear operator that maps real vectors $v$ with coordinates indexed by $\mathbb{F}_2^m$ to real vectors $\widehat{v} = \mathcal{F}_m(v)$ with coordinates indexed by $\mathbb{F}_2^m$ as follows:

$$\widehat{v}_u = \sum_{x \in \mathbb{F}_2^m} (-1)^{u^\mathsf{T} x} v_x.$$

## 5.3 Fast Fourier transformation method

Equivalently, as a matrix relative to the standard basis,

$$\mathcal{F}_m = \bigotimes_{i=1}^{m} \begin{bmatrix} 1 & 1 \\ 1 & -1 \end{bmatrix}.$$

The inverse of $\mathcal{F}_m$ is given by $\mathcal{F}_m/2^m$ (Exercise 5.2). The following result shows that $\mathcal{F}_m$ diagonalizes circulant matrices.

**Theorem 5.3** (Diagonalization of circulant matrices) *Let $M$ be $2^m \times 2^m$ circulant matrix with first row $r$. If $\hat{r} = \mathcal{F}_m(r)$, then*

$$M = \mathcal{F}_m \begin{bmatrix} \hat{r}_{0\cdots 00} & & & \\ & \hat{r}_{0\cdots 01} & & \\ & & \ddots & \\ & & & \hat{r}_{1\cdots 11} \end{bmatrix} \mathcal{F}_m^{-1}.$$

*Proof* The $(u, v)$th coordinate of the right-hand side equals

$$\frac{1}{2^m} \sum_{w \in \mathbb{F}_2^m} \hat{r}_w (-1)^{v^\mathsf{T} w + u^\mathsf{T} w} = \frac{1}{2^m} \sum_{w \in \mathbb{F}_2^m} \hat{r}_w (-1)^{w^\mathsf{T}(u+v)}.$$

Substituting $\hat{r}_w = \sum_{x \in \mathbb{F}_2^m} (-1)^{w^\mathsf{T} x} r_x$ yields

$$\frac{1}{2^m} \sum_{w \in \mathbb{F}_2^m} \sum_{x \in \mathbb{F}_2^m} r_x (-1)^{w^\mathsf{T}(u+v+x)} = \sum_{x \in \mathbb{F}_2^m} r_x \delta^0(u+v+x) = r_{u+v}.$$

Since the coordinates of the matrix on the right-hand side are a function of $u + v$, it is a circulant matrix. Furthermore, since $r$ is equal to the first row of $M$, this matrix is equal to $M$. □

Theorem 5.3 immediately gives an efficient algorithm to compute a matrix-vector product $Mv$. First compute $\mathcal{F}_m^{-1}(v)$. This requires $m2^m$ arithmetic operations, as shown in Exercise 2.5. Next, compute the product of a diagonal matrix and $v$. This requires $2^m$ multiplications. Finally, compute the Fourier transformation of the result.

*Example 5.3* Revisiting Example 5.1, first compute the Fourier transformation of the first row of $M$:

$$\hat{r} = \mathcal{F}_3 \begin{bmatrix} -1 \\ -1 \\ 1 \\ -1 \\ 1 \\ 1 \\ -1 \\ 1 \end{bmatrix} = \begin{bmatrix} -1 & -1 & 1 & -1 & 1 & 1 & -1 & 1 \\ -1 & -1 & -1 & 1 & 1 & 1 & 1 & -1 \\ 1 & -1 & -1 & -1 & -1 & 1 & 1 & 1 \\ -1 & 1 & -1 & -1 & 1 & -1 & 1 & 1 \\ 1 & 1 & -1 & 1 & -1 & -1 & 1 & -1 \\ 1 & 1 & 1 & -1 & -1 & -1 & -1 & 1 \\ -1 & 1 & 1 & 1 & 1 & -1 & -1 & -1 \\ 1 & -1 & 1 & 1 & -1 & 1 & -1 & -1 \end{bmatrix} \begin{bmatrix} -1 \\ -1 \\ 1 \\ -1 \\ 1 \\ 1 \\ -1 \\ 1 \end{bmatrix} = \begin{bmatrix} 0 \\ 0 \\ 0 \\ 0 \\ -4 \\ 4 \\ -4 \\ -4 \end{bmatrix}.$$

Next, compute the Fourier transformation of $w$ and multiply the result with a diagonal matrix $D$ that has $\hat{r}$ on its diagonal:

$$D(\mathcal{F}_3 w) = \begin{bmatrix} 0 & & & & & & & \\ & 0 & & & & & & \\ & & 0 & & & & & \\ & & & 0 & & & & \\ & & & & -4 & & & \\ & & & & & 4 & & \\ & & & & & & -4 & \\ & & & & & & & -4 \end{bmatrix} \begin{bmatrix} -8 \\ 0 \\ 10 \\ -6 \\ 34 \\ -2 \\ 8 \\ 4 \end{bmatrix} = \begin{bmatrix} 0 \\ 0 \\ 0 \\ 0 \\ 136 \\ -8 \\ 32 \\ 16 \end{bmatrix}.$$

Computing the inverse Fourier transformation gives the result

$$\hat{c} = \mathcal{F}_3^{-1}(D\mathcal{F}_3 w) = \frac{1}{8}\mathcal{F}_3 \begin{bmatrix} 0 \\ 0 \\ 0 \\ 0 \\ 136 \\ -8 \\ -32 \\ -16 \end{bmatrix} = \begin{bmatrix} -24 \\ -18 \\ -12 \\ -14 \\ 24 \\ 18 \\ 12 \\ 14 \end{bmatrix}.$$

The right-hand side is equal to the result given in Example 5.1. ▷

## 5.4 Historical remarks

The subdivision of key-recovery attacks into three phases was introduced by Matsui in his paper on the experimental cryptanalysis of the DES, using the terms "data counting" (distillation), "key counting" (analysis) and "exhaustive search" (search). This paper also introduced the approach explained in Section 5.2. The fast Fourier transformation method from Section 5.3 was introduced by Collard, Standaert and Quisquater.

## 5.5 References

Collard, Baudoin, F -X Standaert, and Jean-Jacques Quisquater (2007). "Improving the Time Complexity of Matsui's Linear Cryptanalysis." In: *Information Security and Cryptology-ICISC 2007: 10th International Conference, Seoul, Korea, November 29–30, 2007. Proceedings 10*. Springer, pp. 77–88.

Matsui, Mitsuru (Aug. 1994b). "The First Experimental Cryptanalysis of the Data Encryption Standard." In: *CRYPTO'94*. Ed. by Yvo Desmedt. Vol. 839. LNCS. Springer, Berlin, Heidelberg, pp. 1–11. DOI: 10.1007/3-540-48658-5_1.

## 5.6 Exercises

### Exercise 5.1

In Example 5.1, it was observed that the incorrect key 100 corresponds to a large empirical correlation.

1. Explain this observation and show that the correlation is exactly 9/32.
2. Suppose that the correct value of the three key bits is not 000. Will there still be incorrect keys with large correlation? Which ones?

### Exercise 5.2

Show that the inverse of $\mathcal{F}_m$ is equal to $\mathcal{F}_m/2^m$.

### Exercise 5.3

The convolution $u \circledast v$ of two vectors $u$ and $v$ of length $2^m$, with coordinates indexed by elements of $\mathbb{F}_2^m$, is defined as

$$(u \circledast v)_y = \sum_{x \in \mathbb{F}_2^m} u_x v_{x+y}.$$

1. Let $M$ be a circulant matrix with first row $u$. Show that $Mv = u \circledast v$.
2. Show that $\mathcal{F}_m(u \circledast v) = \mathcal{F}_m(u) \odot \mathcal{F}_m(v)$, where $\odot$ is the coordinate-wise product.

# 6
# Multiple linear cryptanalysis

If more than one good linear approximation is available, it is natural to try to exploit all of them simultaneously. This is called multiple linear cryptanalysis. The first part of this chapter discusses multiple linear cryptanalysis in general. The second part focuses on the special case with a set of masks that forms a vector space, which is called multidimensional linear cryptanalysis.

## 6.1 Multiple linear cryptanalysis

The idea of multiple linear cryptanalysis is to use more than one linear approximation of a function $F\colon \mathbb{F}_2^n \to \mathbb{F}_2^m$.

### 6.1.1 Multiple linear approximations

A *multiple linear approximation* of a function $F\colon \mathbb{F}_2^n \to \mathbb{F}_2^m$ is a set $\Lambda \subseteq \mathbb{F}_2^n \times \mathbb{F}_2^m$ of pairs of input and output masks. Every pair $(u, v)$ in $\Lambda$ is a linear approximation of F with correlation $C_{v,u}^{\mathsf{F}}$. The *capacity* of $\Lambda$ is

$$\mathsf{Cap}(\Lambda) = \sum_{\substack{(u,v)\in\Lambda \\ (u,v)\neq(0,0)}} \left(C_{v,u}^{\mathsf{F}}\right)^2.$$

The reason for defining this quantity is that it determines the best possible data-complexity of a multiple linear attack based on $\Lambda$. This is discussed below, and proven in Chapter 7.

To build a distinguisher using a multiple linear approximation $\Lambda$, estimate each of the correlations $C_{v,u}^{\mathsf{F}}$ with $(u, v)$ in $\Lambda$. The problem then comes down to hypothesis testing with multivariate distributions, but it is shown below that this can often be reduced to the univariate case.

As long as the number of approximations is not too large compared to the number of samples used to estimate the correlations, the multivariate central limit theorem (Theorem A.2 in Appendix A) suggests that the joint distribution of the estimated correlations will be approximately multivariate normal. Compared to the univariate setting, there is a potential difficulty here: the estimators for different linear approximations are not necessarily independent because they are based on the same plaintext-ciphertext pairs. The multivariate normal case is still manageable, since the covariance matrix captures all dependencies. In fact, the following result shows that the noncentral covariances are typically negligible.

**Theorem 6.1** *Let $(u_1, v_1)$ and $(u_2, v_2)$ be linear approximations of a function F, with empirical correlations $\widehat{c}_1$ and $\widehat{c}_2$, respectively. If $\widehat{c}_1$ and $\widehat{c}_2$ are estimated using the same $q$ plaintext-ciphertext pairs with independent and uniform random plaintexts, then their covariance is equal to*

$$\mathsf{Cov}(\widehat{c}_1, \widehat{c}_2) = \left(C^{\mathsf{F}}_{v_1+v_2, u_1+u_2} - C^{\mathsf{F}}_{v_1, u_1} C^{\mathsf{F}}_{v_2, u_2}\right)/q \, .$$

*Proof* The covariance between $\widehat{c}_1$ and $\widehat{c}_2$ is equal to

$$\mathsf{Cov}(\widehat{c}_1, \widehat{c}_2) = \mathbb{E}(\widehat{c}_1 \widehat{c}_2) - \mathbb{E}(\widehat{c}_1)\mathbb{E}(\widehat{c}_2) = \mathbb{E}(\widehat{c}_1 \widehat{c}_2) - C^{\mathsf{F}}_{v_1, u_1} C^{\mathsf{F}}_{v_2, u_2} \, .$$

Let $(\mathbf{x}_1, \mathbf{y}_1), \ldots, (\mathbf{x}_q, \mathbf{y}_q)$ denote the plaintext-ciphertext pairs used to compute $\widehat{c}_1$ and $\widehat{c}_2$. The first term above can be expanded as

$$\mathbb{E}(\widehat{c}_1 \widehat{c}_2) = \frac{1}{q^2} \sum_{i=1}^{q} \sum_{j=1}^{q} \mathbb{E}\left((-1)^{v_1^\mathsf{T} \mathbf{y}_i + u_1^\mathsf{T} \mathbf{x}_i + v_2^\mathsf{T} \mathbf{y}_j + u_2^\mathsf{T} \mathbf{x}_j}\right)$$

$$= \frac{1}{q^2} \sum_{i=1}^{q} \mathbb{E}\left((-1)^{(v_1+v_2)^\mathsf{T} \mathbf{y}_i + (u_1+u_2)^\mathsf{T} \mathbf{x}_i}\right) + \frac{q(q-1)}{q^2} C^{\mathsf{F}}_{v_1, u_1} C^{\mathsf{F}}_{v_2, u_2}$$

$$= \frac{1}{q} C^{\mathsf{F}}_{v_1+v_2, u_1+u_2} + \frac{q-1}{q} C^{\mathsf{F}}_{v_1, u_1} C^{\mathsf{F}}_{v_2, u_2} \, .$$

Subtracting $C^{\mathsf{F}}_{v_1, u_1} C^{\mathsf{F}}_{v_2, u_2}$ from the above yields the result. □

Theorem 6.1 has two important consequences. The first is that the covariance matrix is not completely determined by the correlations of linear approximations in $\Lambda$, unless $\Lambda$ is closed under addition. The second consequence is that the covariance between the estimators of different correlations is often negligible in practice. The reason is that unless F has exceptionally strong linear approximations, the covariances are much smaller than the variances of the individual estimators. Indeed, from Theorem 6.1, the variances are approximately $1/q$, whereas the covariances are approximately $c/q$, with $c$ the correlation of a linear approximation of F. The covariance matrix of the estimators is well approximated by a diagonal matrix as long as $c$ is much smaller than $1/\sqrt{|\Lambda|}$ (typical case) or $1/|\Lambda|$ (worst case).

*Example 6.1*   Let $\Lambda = \{(u_1, v_1), (u_2, v_2)\}$, with $(u_1, v_1) =$ (000000001, 000010000) and $(u_2, v_2) =$ (000000110, 000001000), be a multiple linear approximation of the example cipher from Section 1.1. By Example 2.3, $(u_1, v_1)$ has correlation

$$(-1)^{\kappa_1}/8 \left(1 + (-1)^{\kappa_2}/2\right)\left(1 + (-1)^{\kappa_3}/2\right),$$

with $\kappa_1 = k_0 + k_{10} + k_{22} + k_{31} + 1$, $\kappa_2 = k_{16} + k_{21}$ and $\kappa_3 = k_{13} + k_{23}$. Similarly, the correlation of $(u_2, v_2)$ is equal to (verify this!)

$$(-1)^{\lambda_1}/8 \left(1 + (-1)^{\lambda_2}/2\right),$$

with $\lambda_1 = k_1 + k_2 + k_{16} + k_{21} + k_{30}$ and $\lambda_2 = k_{10} + k_{22} + 1$. As shown by Theorem 6.1, the covariance matrix of $\widehat{\mathbf{c}}$ depends on the correlation of the linear approximation $(u_1 + u_2, v_1 + v_2)$. Analyzing the trails shows that the correlation of this approximation is

$$(-1)^{\mu_1}/8 \left(1 + (-1)^{\mu_2}\right),$$

where $\mu_1 = k_0 + k_1 + k_2 + k_{10} + k_{22} + k_{30} + k_{31} + 1$ and $\mu_2 = \kappa_2 + \lambda_2$.

Choose a key such that $\kappa_1 = \lambda_1 = \mu_1 = \kappa_2 = \lambda_2 = 1$ and $\kappa_3 = \mu_2 = 0$. In this case, the vector of empirical correlations $\widehat{\mathbf{c}}$ has mean

$$\mathbb{E}(\widehat{\mathbf{c}}) = -\begin{bmatrix} 3/32 \\ 1/16 \end{bmatrix}.$$

The covariance matrix $\mathbb{E}\bigl((\widehat{\mathbf{c}} - \mathbb{E}(\widehat{\mathbf{c}}))(\widehat{\mathbf{c}} - \mathbb{E}(\widehat{\mathbf{c}}))^\mathsf{T}\bigr)$ is equal to

$$\frac{1}{q}\begin{bmatrix} 1 & -1/4 \\ -1/4 & 1 \end{bmatrix} - \frac{1}{q}\begin{bmatrix} 9/1024 & 3/512 \\ 3/512 & 1/256 \end{bmatrix}.$$

The second term in the above expression is negligible.

Some sample estimates of the correlations are shown in Figure 6.1. In this case, the covariance is nonnegligible because $(u_1 + u_2, v_1 + v_2)$ was chosen to have exceptionally large absolute correlation. The probability density contours (such as the ellipse in Figure 6.1) are usually more circular.   ▷

### 6.1.2 Distinguishers

Let $\widehat{\mathbf{c}}_1, \ldots \widehat{\mathbf{c}}_{|\Lambda|}$ be the estimated correlations of the linear approximations in a multiple linear approximation $\Lambda$ of a function F. Assume that $\Lambda$ does not contain the trivial linear approximation $(0, 0)$. For the statistical analysis, the

## 6.1 Multiple linear cryptanalysis

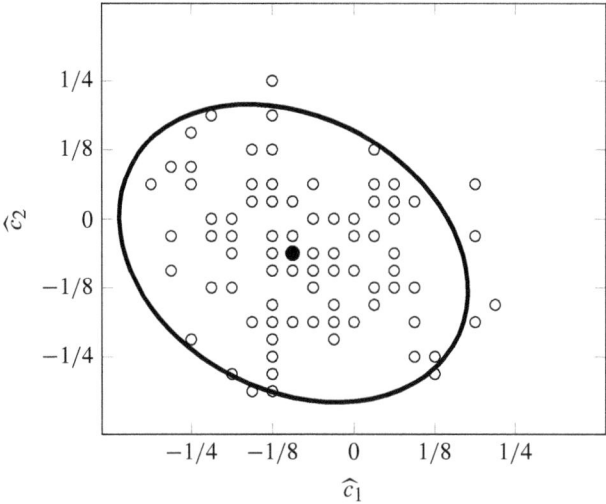

Figure 6.1 Estimated correlations with $q = 64$ samples (100 samples).

simple model from Chapter 4 is used, along with some additional assumptions that are clarified below.

**Known correlations.** Assume that all correlations $C^{\mathsf{F}}_{v,u}$ with $(u,v)$ in $\Lambda$ are known, and that the noncentral covariances are negligible. In principle, building a distinguisher using $\Lambda$ is a multivariate statistics problem. However, we can reduce this problem to a univariate one by using a linear combination of the estimators $\widehat{\mathbf{c}}_1, \ldots, \widehat{\mathbf{c}}_{|\Lambda|}$ as the test statistic:

$$\mathbf{t}_\Lambda = \sum_{i=1}^{|\Lambda|} w_i \, \widehat{\mathbf{c}}_i \,.$$

To keep the variance of $\mathbf{t}_\Lambda$ constant (equal to $1/q$ up to a small error), the weights $w_1, \ldots, w_{|\Lambda|}$ should satisfy $\sum_{i=1}^{|\Lambda|} w_i^2 = 1$. If the samples are uniform random, then the average of $\mathbf{t}_\Lambda$ is equal to zero – provided that $(0,0) \notin \Lambda$. However, if the samples come from the cipher, then the average of $\mathbf{t}_\Lambda$ is equal to

$$\mathbb{E}(\mathbf{t}_\Lambda) = \sum_{i=1}^{|\Lambda|} w_i \, C^{\mathsf{F}}_{v_i, u_i} \,.$$

Theorem 4.1 shows that the data-complexity of a distinguisher based on the test statistic $\mathbf{t}_\Lambda$ is inversely proportional to the square of $\mathbb{E}(\mathbf{t}_\Lambda)$. Hence, it makes sense to maximize $\mathbb{E}(\mathbf{t}_\Lambda)$ while keeping the variance constant. By Exercise 6.1, this is achieved by the choice

$$w_i = \frac{C^F_{v_i,u_i}}{\sqrt{\sum_{i=1}^{|\Lambda|}\left(C^F_{v_i,u_i}\right)^2}}.$$

In this case, the mean is equal to $\sqrt{\mathsf{Cap}(\Lambda)}$. Hence, Theorem 4.1 shows that the data-complexity is proportional to $1/\mathsf{Cap}(\Lambda)$. More precisely, in the simple model and with negligible noncentral covariances, the data-complexity $q$ is

$$q = \frac{\left(\Phi^{-1}(P_S) - \Phi^{-1}(P_F)\right)^2}{\mathsf{Cap}(\Lambda)},$$

where $P_S$ is the success probability and $P_F \leq P_S$ the false-positive probability. It is shown in Chapter 7 that this is essentially optimal.

*Example* 6.2  Consider the multiple linear approximation from Example 6.1. For the same key as before, the capacity is equal to $13/1024$ and the test statistic $\mathbf{t}_\Lambda$ is given by

$$\mathbf{t}_\Lambda = -\frac{3}{\sqrt{13}}\widehat{\mathbf{c}}_1 - \frac{2}{\sqrt{13}}\widehat{\mathbf{c}}_2.$$

A histogram of the distribution of the test statistic is shown in Figure 6.2. ▷

**Unknown correlations.** The correlations of linear approximations typically depend on the key, making it impossible to use the above strategy without guessing key bits. A systematic way to deal with this issue is presented in

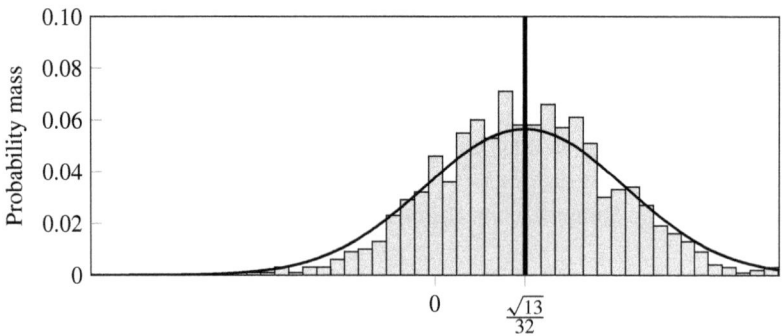

Figure 6.2  Histogram of the test statistic $\mathbf{t}_\Lambda$ for 1000 experiments.

## 6.1 Multiple linear cryptanalysis

Chapter 7. A suboptimal general method and a better method for the case when only the signs of the correlations are unknown are discussed below.

One simple statistical test is based on a linear combination of the squares of the estimated correlations:

$$\mathbf{t}_\Lambda = \sum_{i=1}^{|\Lambda|} w_i \left(\widehat{\mathbf{c}}_i\right)^2,$$

The test statistic $\mathbf{t}_\Lambda$ is not normally distributed, even if all the estimators $\widehat{\mathbf{c}}_1, \ldots, \widehat{\mathbf{c}}_{|\Lambda|}$ are. Nevertheless, if $|\Lambda|$ is large, then the normal distribution will be a good approximation. Hence, to estimate the data-complexity under this assumption, it is enough to determine the mean and variance of $\mathbf{t}_\Lambda$. The following statistical lemma provides this result.

**Lemma 6.2** *Let $\mathbf{x}_1, \ldots, \mathbf{x}_l$ be pairwise uncorrelated random variables with means $\mu_1, \ldots, \mu_l$ and variance $\sigma^2$. The average of $\sum_{i=1}^{l} w_i \mathbf{x}_i^2$ is equal to $\sum_{i=1}^{l} w_i(\sigma^2 + \mu_i^2)$. Furthermore, if $\mathbf{x}_1, \ldots, \mathbf{x}_l$ are normally distributed, then the variance is $2\sigma^2 \sum_{i=1}^{l} w_i^2(\sigma^2 + 2\mu_i^2)$.*

*Proof* The average follows from the observation that the average of $\mathbf{x}_i^2$ is equal to $\mu_i^2 + \sigma^2$, together with $\mathbb{E}(\mathbf{x}_i \mathbf{x}_j) = 0$ for $i \neq j$. If $\mathbf{x}_1, \ldots, \mathbf{x}_n$ are normally distributed and pairwise uncorrelated, then they are independent, so their squares are pairwise uncorrelated too. The variance then follows from

$$\begin{aligned}\mathbb{V}(\mathbf{x}_i^2) &= \mathbb{E}\big((\mathbf{x}_i - \mu_i)^4\big) + 6\mu_i^2 \sigma^2 + \mu_i^4 - \big(\mu_i^2 + \sigma^2\big)^2 \\ &= \mathbb{E}\big((\mathbf{x}_i - \mu_i)^4\big) + 4\mu_i^2 \sigma^2 - \sigma^4 \\ &= 2\sigma^4 + 4\mu_i^2 \sigma^2.\end{aligned}$$

The equality on the first line is the result of a technical calculation, see Exercise 6.2. The last step uses the fact that the fourth moment of the normal distribution $\mathcal{N}(0, 1)$ equals three. This can be shown using integration by parts:

$$\int_{-\infty}^{\infty} x^4 e^{-x^2/2}\, dx = 2\int_{0}^{\infty} x^3 \left(-x\, e^{-x^2/2}\right) dx = 3\int_{-\infty}^{\infty} x^2 e^{-x^2/2}\, dx,$$

where the normalizing factor $1/\sqrt{2\pi}$ was omitted. □

If the samples are uniform random, then Lemma 6.2 with $\mu_1 = \mu_2 = \cdots = 0$ and $\sigma^2 = 1$ shows that the average of $\mathbf{t}_\Lambda$ is equal to $\sum_{i=1}^{|\Lambda|} w_i/q$ and the variance is equal to $2/q^2$ if $\sum_{i=1}^{|\Lambda|} w_i^2 = 1$. If the samples come from the cipher, then the average is equal to (using Lemma 6.2 with $\mu_i = C^F_{v_i, u_i}$ and $\sigma^2 \approx 1/q$)

$$\mathbb{E}(\mathbf{t}_\Lambda) = \frac{1}{q}\sum_{i=1}^{|\Lambda|} w_i + \sum_{i=1}^{|\Lambda|} w_i \left(C^{\mathsf{F}}_{v_i,u_i}\right)^2.$$

The expression for the variance is lengthier, but it is close to $2/q^2$ as long as $q$ is small compared to $1/\left(C^{\mathsf{F}}_{v,u}\right)^2$ for all $(u,v)$ in $\Lambda$. This assumption is reasonable, since otherwise a single approximation would already be sufficient. Translating $\mathbf{t}_\Lambda$ by $\sum_{i=1}^{|\Lambda|} w_i/q$ shows that the hypotheses testing problem amounts to distinguishing between distributions $\mathcal{N}\bigl(\sum_{i=1}^{|\Lambda|} w_i (C^{\mathsf{F}}_{v_i,u_i})^2, 2/q^2\bigr)$ and $\mathcal{N}(0, 2/q^2)$. By Theorem 4.1 with $q^2$ in place of $q$,

$$q = \sqrt{2}\, \frac{\Phi^{-1}(P_{\mathsf{S}}) - \Phi^{-1}(P_{\mathsf{F}})}{\sum_{i=1}^{|\Lambda|} w_i \left(C^{\mathsf{F}}_{v_i,u_i}\right)^2}, \tag{6.1}$$

where $P_{\mathsf{S}}$ is the success probability and $P_{\mathsf{F}} \leq P_{\mathsf{S}}$ the false-positive probability.

This formula relies on the assumptions of the simple model from Section 4.2, as well as the other assumptions mentioned above: the noncentral covariances should be negligible, $|\Lambda|$ should be large and $q$ should be small compared to $1/\left(C^{\mathsf{F}}_{v,u}\right)^2$ for all $(u,v)$ in $\Lambda$. These additional assumptions are often reasonable. However, if the absolute correlations are close to $2^{-n/2}$, then the simple model becomes unreliable. This issue will be discussed in Section 7.3.3. If nothing is known about the correlations at all, then the best one can do is to choose equal weights $w_1 = \cdots = w_{|\Lambda|} = 1/\sqrt{|\Lambda|}$. By (6.1), this leads to a data-complexity of

$$q = \sqrt{2|\Lambda|}\, \frac{\Phi^{-1}(P_{\mathsf{S}}) - \Phi^{-1}(P_{\mathsf{F}})}{\mathrm{Cap}(\Lambda)}.$$

This approach is suboptimal since one almost always has a key-dependent expression for the correlations that should give some idea about their value.

The simplest example is when one knows the absolute values of the correlations, but not the signs. In this case, one should use weights proportional to $\left(C^{\mathsf{F}}_{v_i,u_i}\right)^2$. By (6.1), the data-complexity is proportional to

$$\frac{1}{\sqrt{\sum_{i=1}^{|\Lambda|} \left(C^{\mathsf{F}}_{v_i,u_i}\right)^4}} \leq \frac{\sqrt{|\Lambda|}}{\mathrm{Cap}(\Lambda)}.$$

The upper bound on the right-hand side is matched when the absolute correlations of all the linear approximations in $\Lambda$ are equal. It will be shown in Chapter 7 that in general, this test is still not optimal. When the signs of the correlations of different approximations depend on the same key bits, the data-complexity can often be reduced.

Even if the squared correlations depend on the key as well, the above test can still be used (using some estimate of the squared correlations, such as the average, to determine $w_1, \ldots, w_{|\Lambda|}$), but in this case estimating the data-complexity is more technical because the success probability formula should be averaged over the key as explained in Section 4.2.2. It will be shown in Chapter 7 that, in some cases, one can do better.

## 6.2 Multidimensional linear cryptanalysis

A multidimensional linear approximation is a multiple linear approximation $\Lambda$ such that $\Lambda$ is a vector space over $\mathbb{F}_2$. Since multidimensional linear approximations are a special case of multiple linear approximations, they can be used to build distinguishers in the same way as described in Section 6.1.2. However, the fact that $\Lambda$ is a vector space leads to an interesting alternative description of these distinguishers.

### 6.2.1 Multidimensional linear approximations

A first hint that multidimensional linear approximations are special is given by Theorem 6.1: if $\Lambda$ is a vector space, then the covariance matrix of the estimated correlations is fully determined by the correlations of linear approximations in $\Lambda$. There is a good reason for this: a multidimensional linear approximation is equivalent to a linear projection of plaintext-ciphertext pairs. To make this precise, we need some linear algebra.

Let $U$ be a vector space over $\mathbb{F}_2$. For all subspaces $V$ of $U$, the quotient space $U/V = \{x + V \mid x \in U\}$ is a vector space of dimension $\dim U - \dim V$. The projection $\pi_V : x \mapsto x + V$ is a linear map from $U$ to $U/V$. For all $x$ and $y$ in $U$, these concepts are related as follows:

$$x \equiv y \bmod V \iff \pi_V(x) = \pi_V(y) \iff x - y \in V.$$

Suppose that $U$ is equipped with a symmetric bilinear form $(x, y) \mapsto x \cdot y$ or "dot product." This is a binary operation so that $x \cdot y = y \cdot x$, $0 \cdot x = 0$ and $(x + y) \cdot z = x \cdot z + y \cdot z$. The *orthogonal complement* of a subspace $V$ of $U$ is the vector space

$$V^\perp = \{x \in U \mid x \cdot y = 0 \text{ for all } y \in V\}.$$

Although $\dim V^\perp = \dim U - \dim V$, the subspace $V^\perp$ is not a complement of $V$ in the algebraic sense: it is possible that $V \cap V^\perp \neq \{0\}$.

*Example 6.3* Let $U = \mathbb{F}_2^n \times \mathbb{F}_2^m$ and $V = \Lambda$. The dot product of $(u, v)$ and $(x, y)$ in $U$ is equal to $(u, v) \cdot (x, y) = u^\mathsf{T} x + v^\mathsf{T} y$. Hence,

$$\Lambda^\perp = \left\{(x, y) \in \mathbb{F}_2^n \times \mathbb{F}_2^m \mid u^\mathsf{T} x + v^\mathsf{T} y = 0 \text{ for all } (u, v) \in \Lambda\right\}.$$

Below, the quotient space $(\mathbb{F}_2^n \times \mathbb{F}_2^m)/\Lambda^\perp$ plays an important role. This is a vector space of the same dimension as $\Lambda$. ▷

Using the concepts introduced above, we can state Theorem 6.3. This result is also known as the *Poisson summation formula* in Fourier analysis. This connection will be explained in Chapters 10 and 11.

**Theorem 6.3** *Let $\mathbf{z}$ be a random variable on a vector space $U$ with dot product $(x, y) \mapsto x \cdot y$, and let $V$ be a subspace of $U$. For all $t$ in $U$,*

$$\Pr\left[\mathbf{z} \equiv t \bmod V^\perp\right] = \frac{1}{|V|} \sum_{v \in V} (-1)^{v \cdot t} c_{v \cdot \mathbf{z}},$$

*where $c_{v \cdot \mathbf{z}}$ denotes the correlation of the random variable $v \cdot \mathbf{z}$.*

*Proof* Let $p_{\mathbf{z}}(t) = \Pr[\mathbf{z} = t]$. Recall from Section 2.1 that

$$c_{v \cdot \mathbf{z}} = \sum_{z \in U} (-1)^{v \cdot z} p_{\mathbf{z}}(z).$$

Substituting this into the given sum yields

$$\sum_{v \in V} (-1)^{v \cdot z} c_{v \cdot \mathbf{z}} = \sum_{v \in V} \sum_{z \in U} (-1)^{v \cdot (z+t)} p_{\mathbf{z}}(z) = \sum_{z \in U} p_{\mathbf{z}}(z) \sum_{v \in V} (-1)^{v \cdot (z+t)}.$$

The inner sum can be computed using the same approach that was used in the proof of Theorem 2.4:

$$\sum_{v \in V} (-1)^{v \cdot (z+t)} = \begin{cases} |V| & \text{if } z + t \in V^\perp, \\ 0 & \text{else.} \end{cases}$$

The result then follows from $\Pr[\mathbf{z} \equiv t \bmod V^\perp] = \sum_{z \in t+V^\perp} p_{\mathbf{z}}(z)$. □

Applying Theorem 6.3 to the case of a multidimensional linear approximation gives the following result.

**Corollary 6.4** *Let $\Lambda$ be a multidimensional linear approximation of a function $\mathsf{F}\colon \mathbb{F}_2^n \to \mathbb{F}_2^m$. If $\mathbf{x}$ is a uniform random variable on $\mathbb{F}_2^n$, then*

$$\Pr\left[(\mathbf{x}, \mathsf{F}(\mathbf{x})) \equiv (s, t) \bmod \Lambda^\perp\right] = \frac{1}{|\Lambda|} \sum_{(u, v) \in \Lambda} (-1)^{u^\mathsf{T} s + v^\mathsf{T} t} c^\mathsf{F}_{v, u}$$

*for all $s$ in $\mathbb{F}_2^n$ and $t$ in $\mathbb{F}_2^m$. Furthermore, the above relation is invertible.*

## 6.2 Multidimensional linear cryptanalysis

*Proof* Use Theorem 6.3 with $U = \mathbb{F}_2^n \times \mathbb{F}_2^m$ and $V = \Lambda$, like in Example 6.3. The random variable $\mathbf{z}$ is equal to $(\mathbf{x}, \mathsf{F}(\mathbf{x}))$. The relation in Theorem 6.3 is invertible; finding the inverse relation is Exercise 6.3. □

Corollary 6.4 shows that the correlations of the linear approximations in $\Lambda$ determine the probability distribution of $\pi_{\Lambda^\perp}\big((\mathbf{x}, \mathsf{F}(\mathbf{x}))\big)$. This is a linear projection of the input and output bits. In Chapter 11, it is shown that the relation between the correlations and the probability distribution of $\pi_{\Lambda^\perp}\big((\mathbf{x}, \mathsf{F}(\mathbf{x}))\big)$ is given by the Fourier transformation, and why this is the case.

*Example 6.4* Let $\Lambda = \{(0,0), (u_1, v_1), (u_2, v_2), (u_1 + u_2, v_1 + v_2)\}$ with $(u_1, v_1) = (000000001, 000010000)$ and $(u_2, v_2) = (000000110, 000100000)$. The orthogonal complement $\Lambda^\perp$ consists of all pairs $(x, y)$ such that $x_0 + y_4 = 0$ and $x_1 + x_2 + y_5 = 0$, with $(x_8, \ldots, x_0)$ and $(y_8, \ldots, y_0)$ the coordinates of $x$ and $y$, respectively. Hence, a possible basis for $(\mathbb{F}_2^9 \times \mathbb{F}_2^9)/\Lambda^\perp \cong \mathbb{F}_2^2$ is

$$\{(000000000, 000010000) + \Lambda^\perp, (000000000, 000100000) + \Lambda^\perp\}.$$

Relative to this basis, the coordinates of the projection $(x, y) \bmod \Lambda^\perp$ of $(x, y)$ are given by

$$(x_0 + y_4, x_1 + x_2 + y_5).$$

By Corollary 6.4, the probability that $(\mathbf{x}, \mathsf{F}(\mathbf{x})) \equiv (0, 0) \bmod \Lambda^\perp$ is equal to $1/4\,(1 - 3/32 - 1/16 - 1/4) = 19/128$. ▷

Corollary 6.4 implies that the capacity of a linear approximation can be expressed in a different way. The right-hand side in the following theorem is called the *squared Euclidean imbalance*. Proving this result is Exercise 6.4.

**Corollary 6.5** (Squared Euclidean imbalance) *Let $\Lambda$ be a multidimensional linear approximation of a function $\mathsf{F}\colon \mathbb{F}_2^n \to \mathbb{F}_2^m$. If $\mathbf{x}$ is a uniform random variable on $\mathbb{F}_2^n$, then*

$$\mathsf{Cap}(\Lambda) = |\Lambda| \sum_z \left( \Pr\big[(\mathbf{x}, \mathsf{F}(\mathbf{x})) \equiv z \bmod \Lambda^\perp \big] - \frac{1}{|\Lambda|} \right)^2,$$

*where the sum is over all $z$ in $(\mathbb{F}_2^n \times \mathbb{F}_2^m)/\Lambda^\perp$.*

### 6.2.2 Distinguishers

Since every multidimensional linear approximation is a multiple linear approximation, distinguishers can be obtained as described in Section 6.1.2. However, in light of Corollary 6.4, there is an alternative approach.

Instead of estimating the correlations of linear approximations in $\Lambda$, one can estimate the probability distribution of $\pi_{\Lambda^\perp}((\mathbf{x}, \mathsf{F}(\mathbf{x})))$ for uniform random $\mathbf{x}$.

**Known correlations.** If all correlations are known, then so is the probability distribution of $\pi_{\Lambda^\perp}((\mathbf{x}, \mathsf{F}(\mathbf{x})))$. In this case, one can use the test statistic

$$\mathbf{t}_\Lambda = \sum_{i=1}^{|\Lambda|} w_i \left(\widehat{\mathbf{p}}_i - p_i\right),$$

where $p_1, \ldots, p_{|\Lambda|}$ are the probabilities for the $|\Lambda|$ values in $(\mathbb{F}_2^n \times \mathbb{F}_2^m)/\Lambda^\perp$ and $\widehat{\mathbf{p}}_1, \ldots, \widehat{\mathbf{p}}_{|\Lambda|}$ their estimates. In Exercise 6.5, you will show that choosing optimal weights $w_1, \ldots, w_{|\Lambda|}$ results in a distinguisher with data-complexity inversely proportional to the squared Euclidean imbalance. By Corollary 6.5, the squared Euclidean imbalance is equal to $\mathsf{Cap}(\Lambda)$, so the data-complexity is the same as in Section 6.1.2.

**Unknown correlations.** If the correlations are not known, then a popular approach is to use *Pearsons's* $\chi^2$ *test*. This test is based on the test statistic

$$\mathbf{t}_\Lambda = \sum_{i=1}^{|\Lambda|} \frac{\left(\widehat{\mathbf{p}}_i - 1/|\Lambda|\right)^2}{1/|\Lambda|}.$$

In Exercise 6.6, you will show that the data-complexity of this test is proportional to $\sqrt{|\Lambda|}/\mathsf{Cap}(\Lambda)$. This is same as for the test with equal weights from Section 6.1.2, but worse than the test for unknown correlations with known absolute value.

### 6.2.3 Chosen plaintext attacks

So far, the input to the primitive was assumed to be uniform random on $\mathbb{F}_2^n$. Another consequence of Corollary 6.4 is that multidimensional linear approximations say something about the output when the input is uniform random on an affine subspace of $\mathbb{F}_2^n$. In some cases, this observation is useful to reduce the data-complexity.

If $\Lambda = \Lambda_{\mathsf{in}} \oplus \Lambda_{\mathsf{out}}$, then Corollary 6.4 takes the following form.

**Corollary 6.6** *Let $\Lambda$ be a multidimensional linear approximation of a function $\mathsf{F} \colon \mathbb{F}_2^n \to \mathbb{F}_2^m$. Suppose that $\Lambda = \Lambda_{\mathsf{in}} \oplus \Lambda_{\mathsf{out}}$ with $\Lambda_{\mathsf{in}} \subseteq \mathbb{F}_2^n$ and $\Lambda_{\mathsf{out}} \subseteq \mathbb{F}_2^m$. If $\mathbf{x}$ is a uniform random variable on $s + \Lambda_{\mathsf{in}}^\perp$, then*

$$\Pr\left[\mathsf{F}(\mathbf{x}) \equiv t \bmod \Lambda_{\mathsf{out}}^\perp\right] = \frac{1}{|\Lambda_{\mathsf{out}}|} \sum_{(u,v) \in \Lambda} (-1)^{u^\mathsf{T} s + v^\mathsf{T} t} \, C_{v,u}^{\mathsf{F}},$$

*for all $t$ in $\mathbb{F}_2^m$.*

*Proof* If $\Lambda = \Lambda_{\text{in}} \oplus \Lambda_{\text{out}}$, then also $\Lambda^\perp = \Lambda_{\text{in}}^\perp \oplus \Lambda_{\text{out}}^\perp$. Hence, if $\mathbf{x}'$ is uniform random on $\mathbb{F}_2^n$ and $\mathbf{x}$ is uniform random on $s + \Lambda_{\text{in}}^\perp$, then

$$\Pr\left[(\mathbf{x}', \mathsf{F}(\mathbf{x}')) \equiv (s, t) \bmod \Lambda^\perp\right] = \Pr\left[\mathbf{x}' \equiv s \bmod \Lambda_{\text{in}}^\perp \wedge \mathsf{F}(\mathbf{x}') \equiv t \bmod \Lambda_{\text{out}}^\perp\right]$$
$$= \frac{\Pr\left[\mathsf{F}(\mathbf{x}) \equiv t \bmod \Lambda_{\text{out}}^\perp\right]}{|\Lambda_{\text{in}}|},$$

where in the second step the equality $|\Lambda_{\text{in}}^\perp|/2^n = 1/|\Lambda_{\text{in}}|$ was used. Finally, Corollary 6.4 implies

$$\frac{\Pr\left[\mathsf{F}(\mathbf{x}) \equiv t \bmod \Lambda_{\text{out}}^\perp\right]}{|\Lambda_{\text{in}}|} = \frac{1}{|\Lambda|} \sum_{(u,v) \in \Lambda} (-1)^{u^\mathsf{T} s + v^\mathsf{T} t} C_{v,u}^\mathsf{F}.$$

Multiplying by $|\Lambda_{\text{in}}|$ yields the result, since $|\Lambda_{\text{in}}||\Lambda_{\text{out}}| = |\Lambda|$. □

Using Corollary 6.6, one can set up a chosen-plaintext distinguisher by sampling plaintexts from an affine subspace of $\mathbb{F}_2^n$ and estimating the distribution of a projection of the ciphertexts. Such attacks are also called *statistical saturation attacks*. This terminology will be explained in Chapter 9.

If the correlations, and hence the probability distributions, are known, then the data-complexity of the statistical test from Section 6.2.2 is inversely proportional to the squared Euclidean imbalance, which is equal to

$$|\Lambda_{\text{out}}| \sum_t \left(\Pr\left[\mathsf{F}(\mathbf{x}) \equiv t \bmod \Lambda_{\text{out}}^\perp\right] - \frac{1}{|\Lambda_{\text{out}}|}\right)^2,$$

where the sum is over all $t$ in $\mathbb{F}_2^m/\Lambda_{\text{out}}^\perp$. Although this quantity depends on the choice of the coset $s + \Lambda_{\text{in}}^\perp$, its average is $\mathsf{Cap}(\Lambda)$ for a uniform random choice of $s$. Hence, when all correlations are known, the data-complexity is typically not improved.[1]

However, when the correlations are not known, the data-complexity of the statistical test from Section 6.2.2 is $\sqrt{|\Lambda_{\text{out}}|}/\mathsf{Cap}(\Lambda)$ rather than $\sqrt{|\Lambda|}/\mathsf{Cap}(\Lambda)$. Hence, using chosen plaintexts leads to a data-complexity that is lower by a factor $\sqrt{|\Lambda|/|\Lambda_{\text{out}}|} = \sqrt{|\Lambda_{\text{in}}|}$.

## 6.3 Closing remarks

Before ending our discussion of multiple linear cryptanalysis, some comments about issues that we have ignored are in order.

---

[1] This summary ignores gains from using sampling without replacement.

### 6.3.1 Key recovery

An easy way to apply the key-recovery methods from Chapter 5 to multiple linear cryptanalysis is to repeat the usual algorithms $|\Lambda|$ times; once for every approximation. This approach has the downside that the time-complexity of the analysis phase is $|\Lambda|$ times higher. Hence, although multiple linear cryptanalysis reduces the data-complexity, it may not help improve the time-complexity of a key-recovery attack.

However, one can often do better. It is easy to see, for instance, that improvements are possible when some of the approximations share the same input (or output) mask. There is a systematic approach to this problem, but describing it requires a better understanding of multiple linear cryptanalysis. This will be clarified in Chapter 11.

Finally, the bidirectional key-recovery algorithms from Chapter 5 cannot be combined with the use of chosen plaintexts (or, for that matter, chosen ciphertexts), like in Section 6.2.3. To ensure that plaintexts have the correct structure, additional samples are necessary in general. However, there are interesting exceptions, such as when using an affine subspace of plaintexts together with partial encryption for a key-addition layer.

### 6.3.2 Finding suitable linear approximations

In this chapter, and in Section 6.1 in particular, much attention was devoted to statistical aspects of multiple linear cryptanalysis. In other words, we discussed how to use multiple linear approximations rather than how to find them.

In principle, the methods from Chapters 2 and 3 are sufficient to find suitable linear approximations. However, the most powerful multiple linear attacks construct multiple linear approximations round by round. This too is something to revisit after reading Chapter 11.

## 6.4 Historical remarks

Multiple linear cryptanalysis was proposed by Kaliski and Robshaw. Biryukov, De Cannière and Quisquater analyzed the statistics of multiple linear cryptanalysis when the correlations are known and only their sign depends on the key. An analysis applicable to the case with unknown correlations can be found in the work of Blondeau and Nyberg.

Multidimensional linear cryptanalysis was introduced by Hermelin, Cho and Nyberg. As explained in Section 6.2, the main interest of multidimensional linear cryptanalysis is the relation to distributions of linear projections of the plaintexts and ciphertexts in both the known- (Corollary 6.4) and chosen-plaintext (Corollary 6.6) settings. The use of Pearson's $\chi^2$-test in cryptanalysis predates multidimensional linear approximations and was first proposed by Vaudenay.

## 6.5 References

Biryukov, Alex, Christophe De Cannière, and Michaël Quisquater (2004). "On Multiple Linear Approximations." In: *Advances in Cryptology – CRYPTO 2004, 24th Annual International Cryptology Conference, Santa Barbara, California, USA, August 15–19, 2004, Proceedings.* Ed. by Matthew K. Franklin. Vol. 3152. LNCS. Springer, Berlin, Heidelberg, pp. 1–22. DOI: 10.1007/978-3-540-28628-8\_1. URL: https://doi.org/10.1007/978-3-540-28628-8%5C_1.

Blondeau, Céline and Kaisa Nyberg (2017). "Joint Data and Key Distribution of Simple, Multiple, and Multidimensional Linear Cryptanalysis Test Statistic and Its Impact to Data Complexity." In: *Designs, Codes and Cryptography* 82, pp. 319–349.

Hermelin, Miia, Joo Yeon Cho, and Kaisa Nyberg (2008). "Multidimensional Linear Cryptanalysis of Reduced Round Serpent." In: *Information Security and Privacy, 13th Australasian Conference, ACISP 2008, Wollongong, Australia, July 7–9, 2008, Proceedings.* Ed. by Yi Mu, Willy Susilo, and Jennifer Seberry. Vol. 5107. LNCS. Springer, Berlin, Heidelberg, pp. 203–215. DOI: 10.1007/978-3-540-70500-0\_15. URL: https://doi.org/10.1007/978-3-540-70500-0%5C_15.

Kaliski Jr., Burton S. and Matthew J. B. Robshaw (Aug. 1994). "Linear Cryptanalysis Using Multiple Approximations." In: CRYPTO'94. Ed. by Yvo Desmedt. Vol. 839. LNCS. Springer, Berlin, Heidelberg, pp. 26–39. DOI: 10.1007/3-540-48658-5_4.

Vaudenay, Serge (1996a). "An Experiment on DES Statistical Cryptanalysis." In: *CCS '96, Proceedings of the 3rd ACM Conference on Computer and Communications Security, New Delhi, India, March 14–16, 1996.* Ed. by Li Gong and Jacques Stearn. ACM, New York, pp. 139–147. DOI: 10.1145/238168.238206. URL: https://doi.org/10.1145/238168.238206.

## 6.6 Exercises

### Exercise 6.1

Let $\mu_1, \ldots, \mu_l$ be real numbers, not all zero. Find weights $w_1, \ldots, w_l$ with $\sum_{i=1}^{l} w_i^2 = 1$ such that $\sum_{i=1}^{l} w_i \mu_i$ is maximal. What is the maximum value?

### Exercise 6.2

Let $\mathbf{x}$ be a random variable with mean $\mu$ and variance $\sigma^2$.

1. Prove that if $\mathbf{x}$ is a symmetric random variable, i.e., $-\mathbf{x}$ and $\mathbf{x}$ have the same distribution, then

$$\mathbb{E}(\mathbf{x}^4) = \mathbb{E}((\mathbf{x} - \mu)^4) + 6\mu^2 \sigma^2 + \mu^4.$$

2. Use this result to complete the proof of Lemma 6.2.

### Exercise 6.3

Find the inverse relation of Theorem 6.3.

### Exercise 6.4

Prove Corollary 6.5: the squared Euclidean imbalance of a multidimensional linear approximation $\Lambda$ is equal to its capacity.

### Exercise 6.5

Show that if the probabilities $p_1, \ldots, p_{|\Lambda|}$ are known, then the data-complexity of the test from Section 6.2.2 is proportional to the reciprocal of the squared Euclidean imbalance.

### * Exercise 6.6

Analyze the data-complexity of distinguishers based on Pearson's $\chi^2$ test in terms of the squared Euclidean imbalance. Rely on approximations where appropriate.

In terms of linear cryptanalysis, explain why the $\chi^2$ test-statistic has $|\Lambda| - 1$ degrees of freedom for $|\Lambda|$ categories.

### Exercise 6.7

A block cipher $\mathsf{E}_k : \mathbb{F}_2^{128} \to \mathbb{F}_2^{128}$ has been analyzed using linear cryptanalysis, resulting in the following estimates for the correlations of three dominant linear approximations:

$$C^{E_k}_{01100\cdots0,\,10\cdots00} \approx (-1)^{k_1} \frac{1}{2} + (-1)^{k_2} \frac{1}{4},$$

$$C^{E_k}_{10000\cdots0,\,10\cdots00} \approx (-1)^{k_1} \frac{1}{4} + (-1)^{k_3} \frac{1}{4},$$

$$C^{E_k}_{11100\cdots0,\,10\cdots00} \approx (-1)^{k_1+k_3} \frac{1}{4}.$$

To verify the analysis, experiments were conducted for two different values of the key. In each experiment, a histogram of the values taken by the first three output bits is computed from a sample (with replacement) of 256 plaintexts of the form "$0****\cdots*$," where the $*$ positions are sampled independently and uniformly at random from $\{0, 1\}$. The results are shown in Figure 6.3.

1. Explain why in both Figure 6.3a and 6.3b, the number of occurrences of 100 and 111 is nearly equal.
2. What is the most likely value of the key bits $k_1$, $k_2$ and $k_3$ in the experiment of Figure 6.3a?
3. Find values of $k_1$, $k_2$ and $k_3$ so that the first three ciphertext bits are (almost) never 000.
4. Sketch the most likely histogram for an experiment based on the same key as in Figure 6.3b, but based on a sample (with replacement) of 256 plaintexts of the form "$1**\cdots*$."

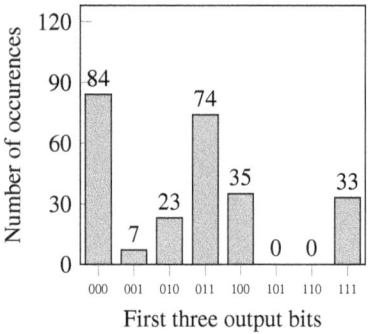

(a) Histogram for the first key.

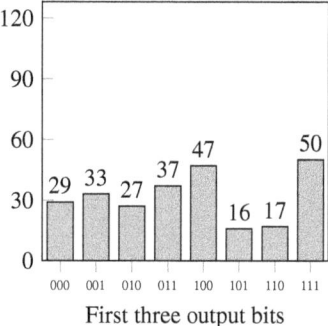

(b) Histogram for the second key.

Figure 6.3 Number of occurences of each value of the first three ciphertext bits in the sample.

## * Exercise 6.8

The following questions lead up to a generic attack on the Lai–Massey construction, which is illustrated for $n = 128$ bits in Figure 6.4. A solution to the first four questions violates the security claims of the tweakable block cipher SPC.[2] The last question is more open-ended.

Suppose the round functions $F_1, F_2, \ldots$ are independent uniform random functions. Random functions are investigated in more detail in Chapter 7. For this exercise, you can assume that every nontrivial linear approximation over an $m$-bit random function has correlation $\pm 2^{-m/2}$ with a uniform random sign.

1. Find a linear trail through three rounds of the $n$-bit Lai–Massey construction with correlation approximately $\pm 2^{-n/4}$.
2. Describe a three-round multidimensional linear distinguisher using approximately $2^{n/4}$ data. Relate it to a $\chi^2$-distinguisher.
3. Use chosen plaintexts to extend your distinguisher to four rounds, using the same amount of data. Something is unusual about this distinguisher, why would it work anyway for a cipher such as SPC?
4. Derive a partial message-recovery attack. That is, obtain *some* new information about plaintexts. You can assume that the plaintexts are already partially known to the adversary.
5. Denote the first Lai–Massey round function by $F_1$. Find a method to obtain the output of $F_1$ for chosen inputs using a similar amount of data as the partial message-recovery attack.
6. In SPC, the function $F_1$ is defined based on a cryptographic function called SipHash-1-2. Find a chosen-tweak and chosen-plaintext key-recovery attack for $F_1$. Deduce a key-recovery attack on full-round SPC.

   *Hint: You may need more than linear cryptanalysis to pull this off.*

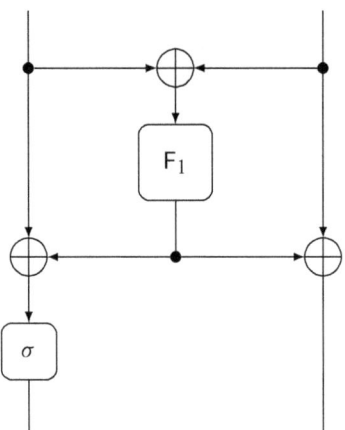

Figure 6.4 One round of the Lai–Massey structure. The function $\sigma : \mathbb{F}_2^{64} \to \mathbb{F}_2^{64}$ is defined by $\sigma(x_1 \| x_2) = x_2 \| (x_1 + x_2)$ with $x_1$ and $x_2$ in $\mathbb{F}_2^{32}$.

---

[2] https://github.com/veorq/spc

# 7
# Optimal statistical testing

In the previous chapters, and in Chapters 4 and 6 in particular, we already encountered methods for testing hypotheses. We used these statistical tests to determine if a given empirical correlation corresponds to the real key, or to an incorrect key. This chapter takes a more systematic look at statistical testing and derives methods that are – in some particular sense – best possible.

## 7.1 Probability measures

Most of the results in this chapter are applicable to both discrete and continuous probability distributions. To avoid repetition, it is convenient to use the language of measure theory. No familiarity with this material is necessary to follow this chapter, provided that the following comments are kept in mind.

A measure space is a triple $(X, \mathfrak{S}, \mu)$ with $X$ a nonempty set, $\mathfrak{S}$ a set of subsets of $X$ that contains $X$ and is closed under complements and countable unions and $\mu$ a measure. A measure is a function $\mu \colon \mathfrak{S} \to \mathbb{R}$ that takes nonnegative values, satisfies $\mu(\emptyset) = 0$ and is *countably additive*: $\mu(\bigcup_{i=1}^{\infty} A_i) = \sum_{i=1}^{\infty} \mu(A_i)$ for disjoint sets $A_1, A_2, \ldots$ in $\mathfrak{S}$. The elements of $\mathfrak{S}$ are called measurable sets.

A real-valued function $f \colon X \to \mathbb{R}$ is called measurable if the preimages of all intervals of the form $(-\infty, t)$ are measurable. In this case, one can define the integral of $f$ over a measurable set $Y$, denoted by

$$\int_Y f(x)\, \mu(dx).$$

The technical details of the definition of such integrals are deliberately left vague. For the purpose of this chapter, it is enough to understand the two examples given at the end of this section. Occasionally, we will say that two measurable functions are equal almost everywhere (a.e.) if they differ on a set of measure zero.

If a measure $P: \mathfrak{S} \to \mathbb{R}$ satisfies $P(X) = 1$, then it is called a *probability measure*. From the point of view of measure theory, probability distributions are probability measures. If there exists a measurable function $p: X \to \mathbb{R}$ so that for all measurable sets $Y$,

$$P(Y) = \int_Y p(x)\mu(dx),$$

then $P$ is called *absolutely continuous* with respect to $\mu$. The function $p$ is called a *Radon–Nykodym density* of $P$.

*Example 7.1* If $X$ is a finite set and $\mathfrak{S}$ is the set of all subsets of $X$, then $\mu$ can be chosen to be the *counting measure*: $\mu(Y) = |Y|$ for all $Y$ in $\mathfrak{S}$. Every probability distribution $P: \mathfrak{S} \to \mathbb{R}$ is absolutely continuous with respect to the counting measure and

$$P(Y) = \int_Y p(x)\mu(dx) = \sum_{x \in Y} p(x).$$

The Radon–Nykodym density $p(x) = P(\{x\})$ is the familiar probability mass function. ▷

*Example 7.2* For $X = \mathbb{R}$, the construction of $\mathfrak{S}$ is slightly more technical – suffice to say it contains all real intervals $Y = (a,b)$. The standard measure on $\mathbb{R}$ is called the Lebesgue measure and satisfies $\mu(Y) = b - a$. If $P$ is a probability measure, then its cumulative distribution function is $x \mapsto P((-\infty, x])$. If the cumulative distribution function is differentiable, then $P$ has a density:

$$P(Y) = \int_Y p(x)\mu(dx) = \int_a^b p(x)\,dx.$$

The integral on the right is a standard Lebesgue integral on $\mathbb{R}$. ▷

## 7.2 Simple hypotheses

In this section, we consider hypothesis testing with so-called *simple* hypotheses. In this case, one has a sample $x$ and two hypotheses ① and ② about the probability distribution it was sampled from:

**Hypothesis ①:** The observation $x$ was sampled from distribution $P_1$.
**Hypothesis ②:** The observation $x$ was sampled from distribution $P_2$.

## 7.2 Simple hypotheses

A subtle but important difference with what we did in Chapter 4 is that the hypotheses ① and ② must fully specify the distributions $P_1$ and $P_2$. For example, when the hypotheses only specify the averages of $P_1$ and $P_2$, they are *not* simple hypotheses.

In the context of linear key-recovery attacks, the observation $x$ is either a vector of empirical correlations or the empirical distribution of a linear projection of the observed plaintext-ciphertext pairs (multidimensional linear cryptanalysis). Hypothesis ① corresponds to the correct key, whereas ② corresponds to incorrect keys. Hence, these hypotheses are only simple when two important assumptions are made: (i) all correlations are known, and (ii) the samples are uniform random for wrong keys. The second assumption is part of the "simple model" from Chapter 4. Section 7.3 discusses what happens when one of these assumptions fails.

### 7.2.1 Neyman–Pearson theory

Although the title of this chapter refers to *optimal testing*, we have not yet defined what this means. Recall from Chapter 4 that every hypothesis test has a success probability $P_S$ and a false-positive probability $P_F$. The trade-off between $P_S$ and $P_F$ was discussed by Neyman and Pearson, who referred to the probabilities $1 - P_S$ and $P_F$ as error rates of the first and second types, respectively. Both types of errors are typically associated with some cost. In general, this is described by a cost function $f(1 - P_S, P_F)$ that is increasing in both variables. For example, in a linear key-recovery attack, the cost function might be the time- or data-complexity of the attack.

At first sight, the above might suggest that there is no single test that minimizes every cost function $f(1 - P_S, P_F)$. However, Neyman and Pearson showed that there is such a test. More specifically, there is a test that minimizes the false-positive probability for every choice of the success probability. Such a test is called *uniformly most powerful*, because $1 - P_F$ is also called the power of the test.

For every simple hypothesis test, there exits a measurable set $\mathcal{A}$ (the "acceptance region") so that hypothesis ① is accepted when $x \in \mathcal{A}$ and hypothesis ② is accepted otherwise. Hence, $P_S = P_1(\mathcal{A})$ and $P_F = P_2(\mathcal{A})$. The Neyman–Pearson lemma provides sets $\mathcal{A}$ such that $P_F$ is minimal for given $P_S$.

**Theorem 7.1** (Neyman–Pearson lemma)  *Let $P_1$ and $P_2$ be probability measures on a measure space $(X, \mathfrak{S}, \mu)$ with densities $p_1$ and $p_2$ (in the Radon–Nykodym sense). For all real $\tau > 0$, let*

$$\mathcal{A}_\tau = \Big\{ x \in X \mid p_1(x) > \tau p_2(x) \Big\}.$$

*If $\mathcal{B}$ is a measurable set such that $P_1(\mathcal{B}) \geq P_1(\mathcal{A}_\tau)$, then $P_2(\mathcal{B}) \geq P_2(\mathcal{A}_\tau)$.*

*Proof* The definition of $\mathcal{A}_\tau$ implies the following inequalities:

$$P_2(\mathcal{B}\setminus\mathcal{A}_\tau) = \int_{\mathcal{B}\setminus\mathcal{A}_\tau} p_2(x)\,\mu(dx) \geq \frac{1}{\tau}\int_{\mathcal{B}\setminus\mathcal{A}_\tau} p_1(x)\,\mu(dx) = \frac{1}{\tau}P_1(\mathcal{B}\setminus\mathcal{A}_\tau).$$

$$P_1(\mathcal{A}_\tau\setminus\mathcal{B}) = \int_{\mathcal{A}_\tau\setminus\mathcal{B}} p_1(x)\,\mu(dx) \geq \tau\int_{\mathcal{A}_\tau\setminus\mathcal{B}} p_2(x)\,\mu(dx) = \tau\, P_2(\mathcal{A}_\tau\setminus\mathcal{B}).$$

Hence,

$$P_2(\mathcal{B}) = P_2(\mathcal{A}_\tau\cap\mathcal{B}) + P_2(\mathcal{B}\setminus\mathcal{A}_\tau) \geq P_2(\mathcal{A}_\tau\cap\mathcal{B}) + \frac{1}{\tau}P_1(\mathcal{B}\setminus\mathcal{A}_\tau).$$

The condition $P_1(\mathcal{B}) \geq P_1(\mathcal{A}_\tau)$ implies that $P_1(\mathcal{B}\setminus\mathcal{A}_\tau) \geq P_1(\mathcal{A}_\tau\setminus\mathcal{B})$. Substituting this in the right-hand side above yields

$$P_2(\mathcal{B}) \geq P_2(\mathcal{A}_\tau\cap\mathcal{B}) + \frac{1}{\tau}P_1(\mathcal{A}_\tau\setminus\mathcal{B}) \geq P_2(\mathcal{A}_\tau\cap\mathcal{B}) + P_2(\mathcal{A}_\tau\setminus\mathcal{B}) = P_2(\mathcal{A}_\tau).$$

Hence, $P_2(\mathcal{B}) \geq P_2(\mathcal{A}_\tau)$ as claimed. $\square$

The acceptance region $\mathcal{A}_\tau$ defined in Theorem 7.1 corresponds to a hypothesis test that compares the *likelihood-ratio test statistic* $t_{lr}$ with the threshold $\tau$. This test statistic is defined by

$$t_{lr}(x) = \frac{p_1(x)}{p_2(x)}.$$

In practice, it is common to use the logarithm of $t_{lr}$ – this is equivalent because the logarithm is an increasing function. The resulting test statistic is called the *logarithmic likelihood ratio* $t_{llr}$:

$$t_{llr}(x) = \log\frac{p_1(x)}{p_2(x)}.$$

Although Theorem 7.1 shows that the (logarithmic) likelihood-ratio test is uniformly most powerful, it does not give the values of $P_S$ and $P_F$. The following two sections determine these values in two special cases: when $P_1$ and $P_2$ are multivariate normal distributions, and when $P_1$ and $P_2$ are nearly equal.

### 7.2.2 Two multivariate normal distributions

In multiple linear cryptanalysis, the empirical correlations are approximately normally distributed when the number of samples $q$ is large enough. This section investigates the likelihood-ratio test for the case of multivariate normal $P_1$ and $P_2$.

Suppose that $P_1$ and $P_2$ are multivariate normal distributions with means $\mu_1$ and $\mu_2$ respectively, and the same $l\times l$ covariance matrix $\Sigma$. For multiple

## 7.2 Simple hypotheses

linear cryptanalysis, $\mu_1$ is a vector of known correlations and $\mu_2 = 0$ (simple model). It was shown in Theorem 6.1 that $\Sigma \approx I/q$ is often a good approximation in this case. For multidimensional linear attacks based on the empirical probability distribution of a linear projection of the samples, $\mu_1$ contains the true probabilities and $\mu_2 \equiv 1/l$. Note that $P_1$ and $P_2$ are degenerate in the multidimensional case due to the inclusion of the trivial $(0, 0)$ approximation or equivalently because the empirical probabilities sum to one. Omitting the trivial approximation or one of the empirical probabilities resolves this issue.

The probability distributions $P_1$ and $P_2$ have densities satisfying

$$p_i(x) \propto \exp\left(-\frac{1}{2}(x - \mu_i)^\mathsf{T} \Sigma^{-1}(x - \mu_i)\right).$$

Hence, up to a constant factor, the logarithmic likelihood ratio is equal to

$$\begin{aligned}t_{\mathsf{llr}} &= (x - \mu_2)^\mathsf{T} \Sigma^{-1}(x - \mu_2) - (x - \mu_1)^\mathsf{T} \Sigma^{-1}(x - \mu_1) \\ &= 2(\mu_1 - \mu_2)^\mathsf{T} \Sigma^{-1} x + \mu_2^\mathsf{T} \Sigma^{-1} \mu_2 - \mu_1^\mathsf{T} \Sigma^{-1} \mu_1.\end{aligned}$$

Up to translation and rescaling, the test-statistic is $\mathbf{t}_{\mathsf{lda}} = (\mu_1 - \mu_2)^\mathsf{T} \Sigma^{-1} \mathbf{x}$. This can be rewritten as a linear combination of the coordinates of $\mathbf{x}$:

$$\mathbf{t}_{\mathsf{lda}} = \sum_{i=1}^{l} w_i\, \mathbf{x}_i.$$

The mean of this test statistic is $(\mu_1 - \mu_2)^\mathsf{T} \Sigma^{-1} \mu_1$ under hypothesis ① and $(\mu_1 - \mu_2)^\mathsf{T} \Sigma^{-1} \mu_2$ under hypothesis ②. The variance is $(\mu_1 - \mu_2)^\mathsf{T} \Sigma^{-1}(\mu_1 - \mu_2)$ – see Appendix A. The choice $w \propto (\mu_1 - \mu_2)^\mathsf{T} \Sigma^{-1}$ maximizes the difference between the means under hypotheses ① and ② while keeping the variance constant. This is precisely the approach that was used in Section 6.1.2.

The method discussed above is called *linear discriminant analysis* in the statistics literature. It has a simple geometric interpretation: the distributions $P_1$ and $P_2$ are separated by the hyperplane orthogonal to the vector $(\mu_1 - \mu_2)^\mathsf{T} \Sigma^{-1}$. This is illustrated in Figure 7.1 for $\Sigma \propto I$. If the covariance matrices of $P_1$ and $P_2$ are not equal, then the optimal test is *quadratic discriminant analysis* instead.

### 7.2.3 Two distributions that are nearly equal

The analysis in Section 7.2.2 shows that the known-correlation tests from Sections 6.1.2 and 6.2.2 are uniformly most powerful in the simple model with the additional approximations from Section 6.1.2. However, this assumes that the distinguisher is based on a vector of empirical correlations (or empirical

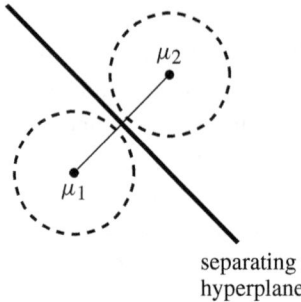

Figure 7.1 Separating $P_1$ and $P_2$ with a line.

probabilities in the multidimensional case). In the multidimensional case, one can go a step further and directly give the distinguisher access to the linear projections $\pi_{\Lambda^\perp}(z_i)$ of the samples $z_1, \ldots, z_q$.

If the samples are independent and identically distributed, then $P_1$ and $P_2$ will be $q$-fold products of distributions $R_1$ and $R_2$. In particular, if $r_1$ and $r_2$ are densities for $R_1$ and $R_2$ respectively, then $p_i(x_1, \ldots, x_q) = \prod_{j=1}^{q} r_i(x_j)$. Hence, the logarithmic likelihood ratio is

$$\mathbf{t}_{\|r} = \log \frac{p_1(\mathbf{x}_1, \ldots, \mathbf{x}_q)}{p_2(\mathbf{x}_1, \ldots, \mathbf{x}_q)} = \sum_{i=1}^{q} \log \frac{r_1(\mathbf{x}_i)}{r_2(\mathbf{x}_i)}.$$

Since the observations $\mathbf{x}_1, \ldots, \mathbf{x}_q$ are independent, the central limit theorem shows that the distribution of $\mathbf{t}_{\|r}/\sqrt{q}$ converges to a normal distribution. Hence, the asymptotic data-complexity of the likelihood-ratio test is determined by Theorem 4.1.

Under hypothesis ①, the average of $\mathbf{t}_{\|r}/q$ is equal to

$$I_{1:2} = \int r_1(x) \log \frac{r_1(x)}{r_2(x)} \mu(dx).$$

Similarly, the average under hypothesis ② is $-I_{2:1}$, where

$$I_{2:1} = \int r_2(x) \log \frac{r_2(x)}{r_1(x)} \mu(dx).$$

Kullback and Leibler[1] referred to $I_{1:2}$ and $I_{2:1}$ as the *mean information of discrimination* between hypotheses ①-② and ②-①, respectively. Furthermore, they defined $J_{12} = I_{1:2} + I_{2:1}$ as the divergence between $R_1$ and $R_2$. Nowadays, $I_{1:2}$ is known as the *Kullback–Leibler divergence* of $R_1$ from $R_2$ and $J_{12}$ is called the *Jeffrys divergence* between $R_1$ and $R_2$.

---

[1] Solomon Kullback and Richard Leibler were both cryptanalysts at the NSA.

## 7.2 Simple hypotheses

Since $J_{12}$ is the difference between the means of the test statistic $\mathbf{t}_{\text{llr}}/q$ under hypotheses ① and ②, it plays an important role in determining the success probability and false-positive probability of the likelihood-ratio test. However, the precise relationship also depends on the variance of $\mathbf{t}_{\text{llr}}/q$. Under hypothesis ①, the variance is equal to $(V_{1:2} - I_{1:2}^2)/q$, where

$$V_{1:2} = \int r_1(x) \left(\log \frac{r_1(x)}{r_2(x)}\right)^2 \mu(dx).$$

Similarly, under hypothesis ②, the variance is $(V_{2:1} - I_{2:1}^2)/q$ with

$$V_{2:1} = \int r_2(x) \left(\log \frac{r_2(x)}{r_1(x)}\right)^2 \mu(dx).$$

If the distributions $R_1$ and $R_2$ are close, then the following lemma simplifies the calculations.

**Lemma 7.2** *Let $r_1$ and $r_2$ be a.e. nonzero probability densities relative to a common probability measure $\mu$ and define $I_{1:2}$, $I_{2:1}$, $V_{1:2}$ and $V_{2:1}$ as above. If $|r_1(x) - r_2(x)| \le \epsilon \min\{|r_1(x)|, |r_2(x)|\}$ a.e., then $I_{2:1} = I_{1:2} + \mathcal{O}(\epsilon^3)$ and*

$$I_{1:2} = \frac{1}{2} \int \frac{(r_1(x) - r_2(x))^2}{r_2(x)} \mu(dx) + \mathcal{O}(\epsilon^3).$$

*Furthermore, $V_{2:1} = V_{1:2} + \mathcal{O}(\epsilon^3)$ and $V_{1:2} = 2\, I_{1:2} + \mathcal{O}(\epsilon^3)$.*

*Proof* If $\varepsilon_{1:2}(x) = (r_2(x) - r_1(x))/r_1(x)$, then

$$I_{1:2} = -\int r_1(x) \log(1 + \varepsilon_{1:2}(x)) \mu(dx)$$

$$= \frac{1}{2}\int r_1(x)\, \varepsilon_{1:2}^2(x)\, \mu(dx) - \underbrace{\int (r_1(x) - r_2(x))\mu(dx)}_{0} + \mathcal{O}(\epsilon^3).$$

The second equality above relies on the Taylor series of $t \mapsto \log(1+t)$ at $t=0$. The second term vanishes because $r_1$ and $r_2$ both integrate to one. Similarly, with $\varepsilon_{2:1}(x) = (r_1(x) - r_2(x))/r_2(x)$,

$$I_{2:1} = \frac{1}{2} \int r_2(x)\, \varepsilon_{2:1}^2(x)\, \mu(dx) + \mathcal{O}(\epsilon^3).$$

The result follows from

$$\frac{(r_1(x) - r_2(x))^2}{r_2(x)} = \frac{(r_1(x) - r_2(x))^2}{r_1(x)} \underbrace{\frac{1}{1 + \varepsilon_{1:2}(x)}}_{1 + \mathcal{O}(\epsilon)} = \frac{(r_1(x) - r_2(x))^2}{r_1(x)} + \mathcal{O}(\epsilon^3) r_1(x).$$

For the second claim, we first show that $V_{1:2} - V_{2:1} = \mathcal{O}(\epsilon^3)$:

$$V_{1:2} - V_{2:1} = \int r_2(x) \underbrace{\varepsilon_{2:1}(x) \log^2\left(1 + \varepsilon_{2:1}(x)\right)}_{\mathcal{O}(\epsilon^3)} \mu(dx) = \mathcal{O}(\epsilon^3).$$

It is now sufficient to show that $V_{2:1} = 2I_{2:1} + \mathcal{O}(\epsilon^3)$. By the Taylor series expansion of $t \mapsto \log(1+t)$ at $t = 0$,

$$\left(\log \frac{r_2(x)}{r_1(x)}\right)^2 = 2\log \frac{r_2(x)}{r_1(x)} + 2\varepsilon_{2:1}(x) + \mathcal{O}(\epsilon^3).$$

Hence, $V_{2:1}$ satisfies

$$V_{2:1} = \int r_2(x) \left(\log \frac{r_2(x)}{r_1(x)}\right)^2 \mu(dx) = 2I_{2:1} + 2\underbrace{\int (r_1(x) - r_2(x))\mu(dx)}_{0} + \mathcal{O}(\epsilon^3).$$

The second term vanishes because $r_1$ and $r_2$ both integrate to one. □

Lemma 7.2 shows that the variance of $t_{\|r}/q$ is the same under hypotheses ① and ②, up to an error $\mathcal{O}(\epsilon^3)$. More precisely, the variance is approximately equal to the difference between the means divided by $q$. Figure 7.2 illustrates the overall situation. By Theorem 4.1, the number of samples $q$ satisfies

$$q = \frac{\left(\Phi^{-1}(P_S) - \Phi^{-1}(P_F)\right)^2}{2 I_{1:2}},$$

assuming that $P_S \geq P_F$. Hence, the data-complexity is inversely proportional to the Kullback–Leibler divergence $I_{1:2}$. For a multidimensional linear approximation $\Lambda \subseteq \mathbb{F}_2^n \times \mathbb{F}_2^m$, the distribution $R_1$ is discrete on $(\mathbb{F}_2^n \times \mathbb{F}_2^m)/\Lambda^\perp$ and $R_2$ is uniform on the same set. For this case, Lemma 7.2 gives the following approximation to $2 I_{1:2}$:

$$2 I_{1:2} = |\Lambda| \sum_z \left(r_1(z) - \frac{1}{|\Lambda|}\right)^2,$$

where the sum is over all $z$ in $(\mathbb{F}_2^n \times \mathbb{F}_2^m)/\Lambda^\perp$. This is precisely the squared Euclidean imbalance, which equals the capacity $\text{Cap}(\Lambda)$ by Corollary 6.5.

## 7.3 Composite hypotheses

Most of the time, the correlations of linear approximations depend on the key. Similarly, for an incorrect guess of the key, the average empirical correlations are (typically) small key-dependent values rather than zero. This means that contrary to Section 7.2, the hypotheses ① and ② do not completely specify the distributions $P_1$ and $P_2$ – they are not simple hypotheses.

### 7.3 Composite hypotheses

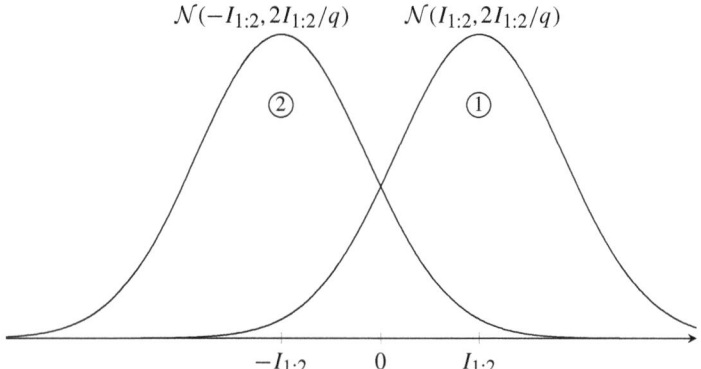

Figure 7.2 Asymptotic distribution of the logarithmic likelihood-ratio test statistic under the hypotheses ① and ②.

In general, given an observation $x$, a composite hypothesis testing problem asks to decide from which one of two families of distributions $x$ was sampled. That is, one has the following two *composite* hypotheses:

**Hypothesis ①:** $x$ was sampled from a distribution in a family $\{P_1^{\theta_1} \mid \theta_1 \in \Theta_1\}$.
**Hypothesis ②:** $x$ was sampled from a distribution in a family $\{P_2^{\theta_2} \mid \theta_2 \in \Theta_2\}$.

One can think of $P_1^{\theta_1}$ and $P_2^{\theta_2}$ as parameterized distributions. The families $\{P_1^{\theta_1} \mid \theta_1 \in \Theta_1\}$ and $\{P_2^{\theta_2} \mid \theta_2 \in \Theta_2\}$ do not need to be disjoint for the problem to make sense, as long as the sets $\Theta_1$ and $\Theta_2$ come with prior probability distributions. For example, if $\Theta_1$ consists of all possible values of the key bits that the correlation depends on, then the prior is the uniform distribution on $\Theta_1$. The prior distributions on $\Theta_1$ and $\Theta_2$ will be called the *right key randomization hypothesis* and *wrong key randomization hypothesis*, respectively. Their choice will be discussed in Sections 7.3.2 and 7.3.3.

The results from Section 7.2 are not applicable for composite hypotheses. In particular, there does not necessarily exist a uniformly most powerful test. Nevertheless, it is possible to find tests that minimize the average $\mathbb{E}f(1 - P_S(\theta_1), P_F(\theta_2))$ of a particular cost function $f$, where the average is with respect to the prior distributions on $\Theta_1$ and $\Theta_2$. This problem is discussed in Section 7.3.1.

#### 7.3.1 Bayes factors

In the absence of a uniformly most powerful test, we can consider tests that are most powerful on average. That is, for all average success probabilities $\mathbb{E}(P_S(\theta_1))$, such a test should minimize the average false-positive probability $\mathbb{E}(P_F(\theta_2))$. This is equivalent to a test between simple hypotheses corresponding to the posterior distributions $P_1$ and $P_2$, whose densities $p_1$ and $p_2$ are given by averaging with respect to the parameters:

$$p_i(x) = \int_{\Theta_i} p_i^\theta(x)\, q_i(\theta)\, \mu(d\theta),$$

where $p_i^\theta$ is the density of $P_i^\theta$ and $q_i$ is the density of the prior distribution.

Using the results from Section 7.2, the likelihood ratio static yields a uniformly most powerful test (on average, as discussed above):

$$t_{lr}(x) = \frac{p_1(x)}{p_2(x)} = \frac{\int p_1^\theta(x)\, q_1(\theta)\, \mu(d\theta)}{\int p_2^\theta(x)\, q_2(\theta)\, \mu(d\theta)}.$$

This quantity is also called the Bayes factor between hypotheses ① and ②.

### 7.3.2 Right key randomization hypothesis

The prior distribution on $\Theta_1$ follows directly from the analysis of the cipher, which results in a key-dependent approximation of each correlation. For example, let $\Lambda$ be a multiple linear approximation. The analysis leads to a set of key classes $\mathcal{K}$, so that the correlations for key class $k$ in $\mathcal{K}$ are given by a known vector $\mu_k$ in $\mathbb{R}^{|\Lambda|}$.

If we assume a uniform random prior distribution for the keys, then the prior probability $f_k$ of every key class $k$ in $\mathcal{K}$ can be computed or, if the key schedule is complex, estimated. Within every key class, the distribution of the empirical correlations is multivariate normal with mean $\mu_k$ and covariance matrix $I/q$ (approximately, as shown by Theorem 6.1). Hence, the probability density $p_1$ is proportional to

$$p_1(x) \propto \sum_{k \in \mathcal{K}} f_k \exp\left(-\frac{q}{2}(x - \mu_k)^\mathsf{T}(x - \mu_k)\right).$$

Such a distribution is known as a multivariate normal mixture. In the simple model, the empirical correlations have mean zero and covariance matrix $I/q$ under hypothesis ②. Hence, the likelihood ratio is proportional to

$$\frac{p_1(x)}{p_2(x)} \propto \sum_{k \in \mathcal{K}} f_k \exp\left(q\, \mu_k^\mathsf{T} x - \frac{q}{2} \mu_k^\mathsf{T} \mu_k\right).$$

In general, there is no "elementary" closed-form expression for the data-complexity of this test. Even if the capacity is (almost) key-independent, the data complexity can be proportional to $1/\mathsf{Cap}(\Lambda)$, $\sqrt{|\Lambda|}/\mathsf{Cap}(\Lambda)$ or something in-between.

7.3 Composite hypotheses     101

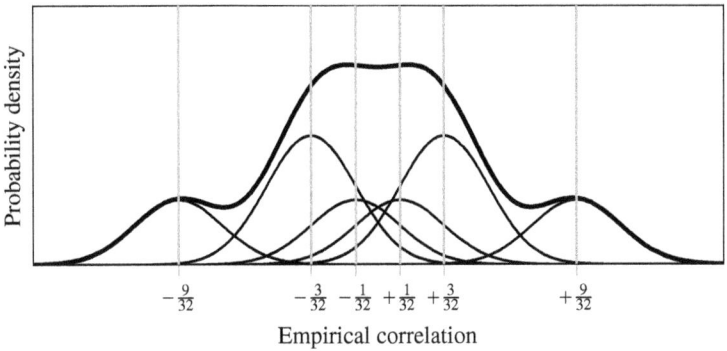

Figure 7.3 Posterior probability density $p_1$ of the empirical correlation.

*Example 7.3* For the linear approximation from Example 2.3 with correlation $(-1)^{\kappa_1}/8\,(1+(-1)^{\kappa_2}/2)(1+(-1)^{\kappa_3}/2)$, the posterior distribution $P_1$ is a mixture of six normal distributions with means $\pm 1/32$, $\pm 3/32$ and $\pm 9/32$. The probability density function $p_1$ of $P_1$ and its mixture components are shown in Figure 7.3 for $q=256$ samples. Up to a constant factor, the likelihood ratio is

$$e^{-\frac{1}{2}q(9/32)^2}\cosh\left(\frac{9qx}{32}\right)+2e^{-\frac{1}{2}q(3/32)^2}\cosh\left(\frac{3qx}{32}\right)+e^{-\frac{1}{2}q(1/32)^2}\cosh\left(\frac{qx}{32}\right).$$

For large $q$, the likelihood ratio is dominated by the term corresponding to correlation $\pm 1/32$. This is due to the exponential factors in each term, the largest of which is $\exp(-q/32^2)$. In other words, because an attack with high average success probability is required to work for most keys, its false-positive probability is mainly determined by the keys with low correlation.  ▷

In special cases, (approximate) closed-form formulas for the data-complexity can be obtained. One such case is when only the signs of the correlations depend on the key. In particular, let $l = |\Lambda|$ and suppose there are positive constants $c_1,\ldots,c_l$ such that

$$\mu_k = \begin{bmatrix}(-1)^{k_1}c_1\\(-1)^{k_2}c_2\\\vdots\\(-1)^{k_l}c_l\end{bmatrix}.$$

Furthermore, assume that $f_k = 1/2^l$ for all $k$. The likelihood ratio is proportional to

$$\frac{p_1(x)}{p_2(x)} \propto \sum_{k \in \mathbb{F}_2^l} \prod_{i=1}^{l} e^{q(-1)^{k_i} c_i x_i} = \prod_{i=1}^{l} \frac{e^{q c_i x_i} + e^{-q c_i x_i}}{2} = \prod_{i=1}^{l} \cosh(q c_i x_i).$$

If $l$ is large, then it can be expected that $q c_i^2$ is small. Since the observations $x_i$ are estimates of the correlation, $q c_i x_i$ is likewise small. Hence, asymptotically as $q c_i x_i \to 0$, the logarithmic likelihood ratio is equal to (up to a constant $C$)

$$\log \frac{p_1(x)}{p_2(x)} + C = \sum_{i=1}^{l} \log \cosh(q c_i x_i) \sim \frac{1}{2} \sum_{i=1}^{l} q^2 c_i^2 x_i^2.$$

That is, the logarithmic likelihood ratio is well approximated by a weighted sum of the squares of the estimated correlations, where the weights themselves are proportional to the squared correlations. This is precisely the test statistic that was used in Section 6.1.2, with data-complexity proportional to

$$\frac{1}{\sqrt{\sum_{i=1}^{l} c_i^4}}.$$

It should be kept in mind that this data-complexity is only optimal when the prior distribution of the key is uniform random.

Finally, note that exact formulas for the correlations are rarely available. Hence, model errors are generally unavoidable. It is possible to take this into account by modifying the prior distribution. For example, one can include a normal error with mean zero. This is useful to counteract overconfidence in the model and can serve as a form of regularization.

### 7.3.3 Wrong key randomization hypothesis

As discussed in Section 7.3.2, hypothesis ① is often composite because correlations are key-dependent. In practice, the correlations for incorrect key guesses are not really zero either. Hence, hypothesis ② should also be composite.

In principle, it is possible to determine approximate key-dependent expressions for the correlations corresponding to incorrect key guesses. However, this requires additional analysis of the cipher – including the outer key-recovery rounds that would not be taken into account in the simple model. The statistical analysis is conceptually the same as in Section 7.3.2.

In the absence of a detailed analysis, there is also a more generic composite refinement of hypothesis ②. If partial encryption and decryption are sufficiently complicated, then one might posit that they resemble random permutations when the key is wrong. This leads to the random permutation model,

## 7.3 Composite hypotheses

which specifies that the prior distribution of correlations under hypothesis ② is the same as for a random permutation.

The distribution of the correlation of a linear approximation of a random function or permutation is given by the following result. The convergence in Theorem 7.3 is fast if $|\Lambda|$ is not too large, so it usually yields good approximations. The proof of this result is obtained in Exercises 7.1 and 7.2.

**Theorem 7.3** *Let $\mathbf{F}$ be a uniform random function or permutation from $\mathbb{F}_2^n$ to $\mathbb{F}_2^m$. Let $\Lambda = \{(u_1, v_1), \ldots, (u_l, v_l)\} \subset \mathbb{F}_2^n \times \mathbb{F}_2^m$ be a multiple linear approximation of $\mathbf{F}$ such that $(0,0) \notin \Lambda$. The probability distribution of the random vector of correlations*

$$\sqrt{2^n} \begin{bmatrix} C^{\mathbf{F}}_{v_1, u_1} \\ C^{\mathbf{F}}_{v_2, u_2} \\ \vdots \\ C^{\mathbf{F}}_{v_l, u_l} \end{bmatrix}$$

*converges to the multivariate normal distribution $\mathcal{N}(0, I)$ as $n \to \infty$.*

For multiple linear cryptanalysis, hypothesis ② then becomes that the empirical correlations have a multivariate normal distribution $\mathcal{N}(\theta, I/q)$. The prior distribution on $\Theta_2$ is $\theta \sim \mathcal{N}(0, I/2^n)$. From the probability density function, it is not difficult to see that this gives a posterior distribution $\mathcal{N}(0, I/q + I/2^n)$. This is illustrated in Figure 7.4 for the univariate case.

The effect of the random permutation model is that it increases the variance of the posterior distribution under hypothesis ②. This also means that the variances of the two posterior distributions are different. For sampling without

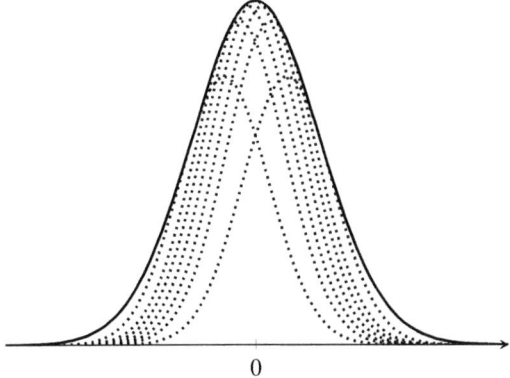

Figure 7.4 Wrong key hypothesis based on the random permutation model.

replacement, the variance under hypothesis ② with known correlation is $1/q\,(1-q/2^n)+1/2^n=1/q$.

Although the refined wrong key hypothesis based on the random permutation model often has a minor impact on the cost estimates of attacks, in particular when $q \ll 2^n$, it leads to important conclusions. Together with Section 7.3.2, it provides a better explanation for the difficulty of using linear approximations with absolute correlation $c$ below $2^{-n/2}$. A naive explanation is that, since the data complexity is proportional to $1/c^2$, there is simply not enough data available. However, using multiple linear approximations does not resolve the issue. A better explanation is that the correlation $c$ is only known up to a modeling error $\varepsilon$ under hypothesis ①, and the refined wrong key hypothesis implies that the modeling error $\varepsilon$ should be less than $2^{-n/2}$. This will play a major role in Chapter 8.

For multiple linear cryptanalysis, the impact of the wrong key hypothesis is particularly important. If $|\Lambda|$ is large, then the analysis can often be simplified because the capacity for wrong keys is close to its mean value $|\Lambda|/2^n$. Based on the discussion in Section 6.1.2, if $q$ is small compared to $1/(C_{v,u}^F)^2$ and the correlations are unknown, then

$$q = \sqrt{2|\Lambda|}\,\frac{\Phi^{-1}(P_S) - \Phi^{-1}(P_F)}{\mathsf{Cap}(\Lambda) - |\Lambda|/2^n},$$

assuming the other assumptions from Section 6.1.2 remain valid.

## 7.4 Optimal key-recovery

In Chapter 1, another approach to key-recovery was mentioned: Matsui's Algorithm 1. This method can be generalized to a classification problem: given a vector of empirical correlations, find the most probable value of the key bits that determine the correlations. There is an optimal solution to this problem, known as the Bayes classifier.

However, there is more to solving this classification problem. A first issue is that it may be impossible to recover the key uniquely. For example, if the correlation is $(-1)^{\kappa_1}/8\,(1+(-1)^{\kappa_2}/2)(1+(-1)^{\kappa_3}/2)$, then swapping the values $\kappa_2$ and $\kappa_3$ always leads to equal likelihoods. This observation leads to a second issue: it is more useful to obtain a list of possible keys rather than a single candidate. Hence, there is a trade-off between the number of key classes and the probability of correct classification.

We do not discuss the optimal way to solve such classification problems here, as that would take us beyond the state of the art. However, to end this section, it is worth pointing out that the "Algorithm 2" approach to key-recovery is in fact also more like a classification than a testing problem. In Sections 7.2 and 7.3, like in Chapter 4, it was assumed that the key-recovery process can be considered as a form of multiple hypothesis testing. However, in this approach the multiple hypothesis tests are not actually statistically independent. The dependencies between estimated correlations for different keys are easier to take into account in the classification framework.

## 7.5 Historical remarks

The theory of Neyman and Pearson for simple hypotheses testing was first applied to linear cryptanalysis by Baignères, Junod and Vaudenay. Their analysis is applicable to multidimensional linear cryptanalysis and is comparable to the discussion in Section 7.2.3.

The generic wrong key randomization hypothesis from Section 7.3.3 was introduced by Bogdanov and Tischhauser. The term "wrong key randomization" was coined by Harpes, Kramer and Massey. For a single linear approximation and for sampling without replacement in particular, the posterior distribution of the test statistic under this hypothesis was discussed by Ashur, Beyne, and Rijmen (2020).

## 7.6 References

Ashur, Tomer, Tim Beyne, and Vincent Rijmen (Apr. 2020). "Revisiting the Wrong-Key-Randomization Hypothesis." In: *Journal of Cryptology* 33.2, pp. 567–594. DOI: 10.1007/s00145-020-09343-2.

Baignères, Thomas, Pascal Junod, and Serge Vaudenay (Dec. 2004). "How Far Can We Go Beyond Linear Cryptanalysis?" In: *ASIACRYPT 2004*. Ed. by Pil Joong Lee. Vol. 3329. LNCS. Springer, Berlin, Heidelberg, pp. 432–450. DOI: 10.1007/978-3-540-30539-2_31.

Bogdanov, Andrey and Elmar Tischhauser (Mar. 2014). "On the Wrong Key Randomisation and Key Equivalence Hypotheses in Matsui's Algorithm 2." In: *FSE 2013*. Ed. by Shiho Moriai. Vol. 8424. LNCS. Springer, Berlin, Heidelberg, pp. 19–38. DOI: 10.1007/978-3-662-43933-3_2.

Harpes, Carlo, Gerhard G. Kramer, and James L. Massey (May 1995). "A Generalization of Linear Cryptanalysis and the Applicability of Matsui's Piling-Up Lemma." In: *EUROCRYPT'95*. Ed. by Louis C. Guillou and Jean-Jacques Quisquater. Vol. 921. LNCS. Springer, Berlin, Heidelberg, pp. 24–38. DOI: 10.1007/3-540-49264-X_3.

Kullback, Solomon and Richard A. Leibler (1951). "On Information and Sufficiency." In: *The Annals of Mathematical Statistics* 22.1, pp. 79–86.

## 7.7 Exercises

### Exercise 7.1

Prove Theorem 7.3 for the case of uniform random functions. Use the multivariate central limit theorem.

### * Exercise 7.2

Prove Theorem 7.3 for the case of uniform random permutations.

### Exercise 7.3

Let $\Lambda$ be a multiple linear approximation consisting of linear approximations with correlations $c_k$ in $\mathbb{R}^{|\Lambda|}$ for every key $k$. Suppose that the prior distribution of $c_\mathbf{k}$ with a uniform random key $\mathbf{k}$ is given by a multivariate normal distribution with mean zero and covariance matrix $\Sigma$. Use the assumptions of the simple model.

1. Show that there exists a linear change of variables of the empirical correlations so that the logarithmic likelihood-ratio test statistic is, up to scaling and translation, equal to a weighted sum of squares.
2. Show that the data-complexity is proportional to $1/\sqrt{\operatorname{Tr}\Sigma^T\Sigma}$.

# 8
# Zero-correlation approximations

Traditionally, linear cryptanalysis exploits linear approximations with atypically high absolute correlation. In this chapter, we discuss instead how linear approximations with correlation zero can be used. This variant of linear cryptanalysis is called *zero-correlation linear cryptanalysis*.

## 8.1 The idea

Loosely speaking, for a key-recovery attack in the style of Matsui's Algorithm 2, it is enough to find a property of the inner part of the cipher that makes it possible to distinguish between the right and wrong key guesses. The simple model from Chapter 4 assumes that the correlations of linear approximations are zero for wrong keys. From the point of view of this simplified model, a linear approximation with correlation zero is not useful. However, as discussed in Chapter 7, correlations are not exactly zero for wrong keys under more accurate wrong key randomization hypotheses such as the random permutation model.

Although the random permutation model implies that linear approximations with correlation zero could be useful *in principle*, some issues need to be addressed. A first problem is finding zero-correlation linear approximations. The difficulty here is that it is not enough that the correlation is small; it must be exactly zero. This issue will be addressed in Section 8.2. Another question is whether or not zero-correlation approximations are "remarkable" enough to be useful as distinguishing properties. The success probability can always be brought up to one by using all possible plaintext-ciphertext pairs to evaluate the correlation – but to filter out enough wrong keys, the false-positive probability must be low as well. After all, by Theorem 7.3, zero is still the most likely value of the correlation for a random permutation.

The reason that low false-positive probabilities can be achieved is related to a somewhat counter-intuitive (at least, to nonspecialists) property of probability distributions. As the number of possible outcomes increases, the probability of every individual outcome – even the most likely one – decreases. This is in contrast to the probability of sufficiently wide intervals of outcomes. Applying this fact to the case of linear cryptanalysis, we observe that for a permutation sampled uniformly at random, most approximations will have a correlation "close to zero," but a correlation equal to zero will rarely occur.

An approximation with correlation zero can be used as a distinguishing property, provided that our estimate for the correlation is accurate enough to distinguish between zero and "close to zero." If only one zero-correlation approximation is available, this leads to a data-complexity that is close to $2^n$ for an $n$-bit function. Sections 8.3 and 8.4 discuss methods to reduce the data-complexity.

Finally, note that in principle every correlation value (or even range of values) could be used as a distinguisher, as long as the correlation is known precisely enough. The value zero is special because it is often easier to show that some linear approximations have correlation zero.

## 8.2 Finding approximations with correlation zero

As shown in Corollary 2.8, the correlation of a linear approximation can be written as the sum of the correlations of linear trails. Although it is sufficient that the sum of the trail correlations is zero, almost all zero-correlation approximations described in the literature have the property that all linear trails have correlation zero.

The existence of linear approximations so that all trails have correlation zero, and the fact that some of them can be found easily, is related to the use of functions with a high linear branch number (see Definition 3.1), and more generally to the use of round functions with a simple structure.

If all linear trails within a linear approximation have correlation zero, then this can be verified using the automated methods from Chapter 3. It suffices to search for linear trails with nonzero correlation without trying to maximize their correlation; if no solutions are found, then the linear approximation has correlation zero. Unfortunately, this approach does not offer much insight into which linear approximations have correlation zero.

Below we discuss the more insightful *miss-in-the-middle* method, that can often be applied by hand. Before discussing the method in general, we give an example for three rounds of the example cipher from Section 1.1.

## 8.2 Finding approximations with correlation zero

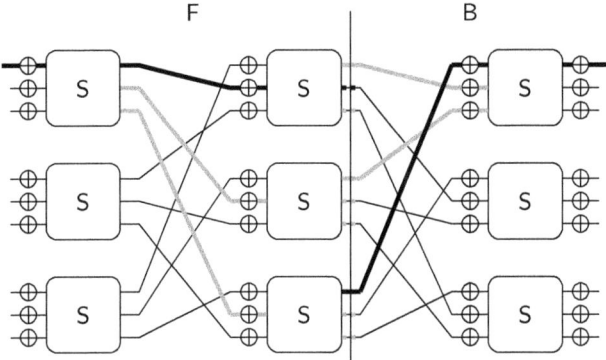

Figure 8.1 Zero-correlation linear approximation using miss-in-the-middle.

*Example* 8.1 (Miss-in-the-middle) Consider three rounds of the example cipher from Section 1.1 as in Figure 8.1. This example shows that the linear approximation $(u, v) = (000000001, 000000001)$ has correlation zero. To do so, we consider all linear approximations with input mask $000000001$ over the first half of the cipher and all linear approximations with output mask $000000001$ over the second half.

Let $\mathsf{E}_k = \mathsf{B} \circ \mathsf{F}$, with F and B as indicated in Figure 8.1. In particular, F consists of the first round and the S-box layer of the second round, and B consists of the bit-permutation of the second round and the S-box layer of the third round. The correlation of $(u, v)$ over $\mathsf{E}_k$ is equal to

$$C^{\mathsf{E}_k}_{v,u} = \sum_{w \in \mathbb{F}_2^9} C^{\mathsf{B}}_{v,w} C^{\mathsf{F}}_{w,u}.$$

Hence, to show that $(u, v)$ is a zero-correlation linear approximation, it is sufficient to show that for all trails $(u, w, v)$, either $C^{\mathsf{F}}_{w,u} = 0$ or $C^{\mathsf{B}}_{v,w} = 0$.

Since row 001 of $C^S$ contains nonzero entries only in columns $ab1$ with $a$ and $b$ in $\mathbb{F}_2$, the condition $C^{\mathsf{B}}_{v,w} \neq 0$ can be rewritten as

$$w \in V = \{00100b00a \mid a, b \in \mathbb{F}_2\}.$$

Similarly, column 001 of $C^S$ has nonzero entries in rows $ab1$ with $a$ and $b$ in $\mathbb{F}_2$. Hence, after the first round, all masks that lead to a nonzero correlation are of the form $0b00a0010$. From the property that column 010 of $C^S$ only has nonzero entries in rows $a1b$ with $a$ and $b$ in $\mathbb{F}_2$, it follows that $C^{\mathsf{F}}_{w,u} \neq 0$ implies

$$w \in U = \{c \parallel a1b \mid a, b \in \mathbb{F}_2, c \in \mathbb{F}_2^6\}.$$

If there exists a linear trail $(u, w, v)$ for $\mathsf{B} \circ \mathsf{F}$ with nonzero correlation, then $w \in U$ and $w \in V$ as argued above. However, $U$ and $V$ are disjoint, so $(u, v)$ is a zero-correlation approximation.

Verify that the same argument works for every mask $v$ with middle three bits equal to zero. For example, (000000001, 000000010), (000000001, 000000100) and (000000001, 000000110) are all zero-correlation linear approximations.

Exercise 8.1 asks you to show that all linear approximations with input mask 000001000 or 000001001 and output mask zero except in the first three bits also have correlation zero. ▷

In general, the miss-in-the-middle technique that was illustrated in Example 8.1 works as follows. Let $\mathsf{E}_k = \mathsf{B} \circ \mathsf{F}$ be a cipher, where $\mathsf{F}$ and $\mathsf{B}$ correspond to an arbitrary decomposition into two parts. For example, $\mathsf{F}$ might consist of the first $r_\mathsf{F}$ rounds of the cipher, and $\mathsf{B}$ of the remaining $r_\mathsf{B}$ rounds. The miss-in-the-middle approach leads to zero-correlation approximations for $r_\mathsf{F} + r_\mathsf{B}$ rounds.

Starting from an input mask $u$, which typically has a particular structure such as having a low Hamming weight, we determine a set of output masks $w$ such that $(u, w)$ might have nonzero correlation over $\mathsf{F}$. This is typically done by propagating masks forward round by round, like in Example 8.1. This results in a set of masks $U$ such that

$$U \supseteq \{w \in \mathbb{F}_2^n \mid C^\mathsf{F}_{w,u} \neq 0\}.$$

Similarly, starting from an output mask $v$, we determine a set of masks $V$ such that

$$V \supseteq \{w \in \mathbb{F}_2^n \mid C^\mathsf{B}_{v,w} \neq 0\}.$$

To construct $V$, the mask $v$ is propagated backwards through $\mathsf{B}$ round by round. The sets $U$ and $V$ are typically described implicitly, such as by patterns or by linear equations, rather than in terms of their elements. The correlation of the linear approximation $(u, v)$ is then given by

$$C^{\mathsf{E}_k}_{v,u} = \sum_{w \in U \cap V} C^\mathsf{B}_{v,w} C^\mathsf{F}_{w,u}.$$

Hence, if $U \cap V$ is empty, then $(u, v)$ is a zero-correlation linear approximation.

The choice of the approximation $(u, v)$ depends on the details of the function. Typically, the structure of $\mathsf{F}$ and $\mathsf{B}$ suggests one or more promising candidates.

## 8.3 Using zero-correlation approximations

This section takes a closer look at how zero-correlation linear approximations can be used as a distinguisher. The analysis assumes that enough plaintext-ciphertext pairs are available to compute the exact correlations of the approximations.

If correlations can be evaluated exactly, then the success probability is equal to one. The first goal of this section is to compute the corresponding false-positive probability.

At first sight, zero-correlation linear cryptanalysis might seem impractical because computing exact correlations appears to require all possible plaintext-ciphertext pairs. The second goal of this section is to show that a smaller number of chosen plaintext-ciphertext pairs are often sufficient when multiple approximations are available.

### 8.3.1 Single approximation

A zero-correlation approximation can be used for key-recovery by using statistical hypothesis testing. Most of the techniques from Sections 4.2 and 5.1 are applicable without adjustments.

As explained in Section 5.1, the block cipher is subdivided into an inner part and an outer part, consisting of the functions $F_l$ and $B_k$. The zero-correlation approximation holds over the inner part. The estimated correlation over the inner part is computed using (5.1):

$$\widehat{c}_{k,l} = \frac{1}{q} \sum_{i=1}^{q} (-1)^{F_l(x_i) + B_k(y_i)}.$$

Unlike in ordinary linear cryptanalysis, we use all possible plaintext-ciphertext pairs to compute $\widehat{c}_{k,l}$. Hence, the "estimate" equals the actual correlation. In particular, $\widehat{c}_{k,l} = 0$ if $k$ and $l$ are the right keys. For incorrect keys, we use the wrong-key-randomization hypothesis based on the random permutation model from Section 7.3.3. In this model, it is unlikely that $\widehat{c}_{k,l} = 0$. The exact probability is given by Corollary 8.2, which is a consequence of the more general Theorem 8.1. This probability is the false-positive probability of the test.

**Theorem 8.1** (Correlation for a random permutation) *Let* $\mathbf{F}$ *be a uniform random permutation on* $\mathbb{F}_2^n$. *The probability that a linear approximation* $(u, v)$ *of* $\mathbf{F}$ *with* $u, v \neq 0$ *has correlation* $4w/2^n - 1$, *is equal to*

$$\Pr\left[C_{v,u}^{\mathbf{F}} = 4w/2^n - 1\right] = \frac{\binom{2^{n-1}}{w}^2}{\binom{2^n}{2^{n-1}}},$$

*assuming that* $w \leq 2^{n-1}$ *is a nonnegative integer.*

*Proof* There are $2^n!$ permutations on $\mathbb{F}_2^n$. To prove the result, it is enough to count the number of permutations so that $(u, v)$ has correlation equal to $4w/2^n - 1$ with $0 \le w \le 2^{n-1}$. To do so, partition $\mathbb{F}_2^n$ into the two subsets $\{x \in \mathbb{F}_2^n \mid u^T x = 0\}$ and $\{x \in \mathbb{F}_2^n \mid u^T x = 1\}$. This results in the following distribution of the input values (like in the proof of Theorem 1.1):

|  | $u^T x = 0$ | $u^T x = 1$ |
|---|---|---|
| $v^T F(x) = 0$ | $w$ | $2^{n-1} - w$ |
| $v^T F(x) = 1$ | $2^{n-1} - w$ | $w$ |

There are $2^{n-1}$ values $x$ such that $u^T x = 0$. For the images $F(x)$, $w$ values must be in the set $\{y \in \mathbb{F}_2^n \mid v^T y = 0\}$ and $2^{n-1} - w$ in its complement. Hence, the number of ways to choose the images is

$$\binom{2^{n-1}}{w}\binom{2^{n-1}}{2^{n-1} - w} = \binom{2^{n-1}}{w}^2.$$

There are $2^{n-1}!$ ways to assign these images to the inputs $x$, so the number of assignments of images $F(x)$ to inputs $x$ with $u^T x = 0$ is

$$2^{n-1}! \binom{2^{n-1}}{w}^2.$$

For the assignments of images $F(x)$ to inputs $x$ such that $u^T x = 1$, the set of $2^{n-1} - w$ images from $\{y \in \mathbb{F}_2^n \mid v^T y = 0\}$ must be the complement of the set of $w$ values that has already been chosen for the case $u^T x = 0$. Similarly, the set of $w$ values from $\{y \in \mathbb{F}_2^n \mid v^T y = 0\}$ has already been determined. Hence, since there are $2^{n-1}!$ ways to assign the images to the inputs, the total number of permutations with correlation $4w/2^n - 1$ is exactly

$$(2^{n-1}!)^2 \binom{2^{n-1}}{w}^2.$$

Dividing by the total number of permutations yields the following probability:

$$\frac{(2^{n-1}!)^2 \binom{2^{n-1}}{w}^2}{2^n!} = \frac{\binom{2^{n-1}}{w}^2}{\binom{2^n}{2^{n-1}}}.$$

This completes the proof. □

## 8.3 Using zero-correlation approximations

**Corollary 8.2** (Zero-correlation for a random permutation)  Let **F** be a uniform random permutation on $\mathbb{F}_2^n$. The probability that a linear approximation $(u, v)$ of **F** with $u, v \neq 0$ has correlation zero, is equal to

$$\Pr\left[C_{v,u}^{\mathsf{F}} = 0\right] = \frac{\binom{2^n-1}{2^{n-2}}^2}{\binom{2^n}{2^{n-1}}} = 2\sqrt{\frac{2}{\pi}}\, 2^{-n/2} + \mathcal{O}\!\left(2^{-3n/2}\right),$$

as $n \to \infty$.

*Proof*  The result follows from Theorem 8.1 by taking $w = 2^{n-2}$. The asymptotic expansion follows from the following estimate:

$$\binom{2N}{N} = \frac{2^{2N}}{\sqrt{\pi}}\left(N^{-1/2} + \mathcal{O}(N^{-3/2})\right),$$

which is a consequence of Stirling's approximation for the factorial.  □

Corollary 8.2 implies that if there are $K$ possible keys, approximately $P_{\mathsf{F}} K \approx 0.8 \times K/2^{n/2}$ of these remain after filtering. Although there are cases with more than $2^{n/2}$ keys, this is usually not a problem because there is often more than one zero-correlation linear approximation available.

The downside of zero-correlation linear cryptanalysis is that, since all possible plaintext-ciphertext pairs are used to evaluate the correlation, the data-complexity is $2^n$. If the outer part of the cipher consists only of one or more final rounds, i.e., $\mathsf{F}_l(x) = u^\mathsf{T} x$ with $u$ the input mask, then by Exercise 1.7 it is sufficient to encrypt all inputs in the set $\{x \in \mathbb{F}_2^n \mid u^\mathsf{T} x = 0\}$. However, a data-complexity of $2^{n-1}$ is still impractical in most cases. Section 8.3.2 shows that if multiple zero-correlation approximations are available, then the data-complexity can be reduced.

### 8.3.2 Multiple approximations

Let $\Lambda$ be a multidimensional linear approximation. If $\Lambda$ has capacity zero, then Corollary 6.6 implies that

$$\Pr\left[(\mathbf{x}, \mathsf{F}(\mathbf{x})) \equiv (s, t) \bmod \Lambda^\perp\right] = \frac{1}{|\Lambda|} \sum_{(u,v) \in \Lambda} (-1)^{u^\mathsf{T} s + v^\mathsf{T} t}\, C_{v,u}^{\mathsf{F}} = \frac{1}{|\Lambda|}.$$

In other words, if $\mathbf{x}$ is uniform random, then so is $(\mathbf{x}, \mathsf{F}(\mathbf{x})) \bmod \Lambda^\perp$. The converse relation also holds: if $(\mathbf{x}, \mathsf{F}(\mathbf{x})) \bmod \Lambda^\perp$ is uniformly distributed, then the nonzero pairs in $\Lambda$ are linear approximations with correlation zero.

This provides an alternative description of multidimensional zero-correlation linear cryptanalysis, but it does not reduce the data-complexity. However, if $\Lambda = \Lambda_{in} \oplus \Lambda_{out}$ with $\Lambda_{in} \subseteq \mathbb{F}_2^n$ and $\Lambda_{out} \subseteq \mathbb{F}_2^n$, then Corollary 6.6 implies that

$$\Pr\left[F(\mathbf{x}) \equiv t \bmod \Lambda_{out}^\perp\right] = \frac{1}{|\Lambda_{out}|} \sum_{(u,v) \in \Lambda} (-1)^{u^T s + v^T t} C_{v,u}^F = \frac{1}{|\Lambda_{out}|},$$

for $\mathbf{x}$ uniform random on $s + \Lambda_{in}^\perp$. This result can be used to reduce the data-complexity.

In particular, it suffices to encrypt a set of the form $s + \Lambda_{in}^\perp$. The distinguishing property then consists of verifying that the number of inputs $x$ such that $F(x) \equiv t \bmod \Lambda_{out}^\perp$ is the same for all values of $t$ in $\mathbb{F}_2^n / \Lambda_{out}^\perp$. Hence, the data-complexity is only $2^n/|\Lambda_{in}|$.

*Example* 8.2  Let $\Lambda = \Lambda_{in} \oplus \Lambda_{out}$ with $\Lambda_{in} = \text{Span}\{000000001, 000001000\}$ and $\Lambda_{out} = \text{Span}\{000000001, 000000010, 000000100\}$. It was shown in Example 8.1 and Exercise 8.1 that $\Lambda$ is a multidimensional zero-correlation linear approximation for three rounds of the example cipher. Hence, if $\mathbf{x}$ is uniform random on $s + \Lambda_{in}^\perp$, then

$$\Pr\left[E_k(\mathbf{x}) \equiv t \bmod \Lambda_{out}^\perp\right] = \frac{1}{8},$$

with $E_k$ three rounds of the example cipher.

The vector space $\Lambda_{in}^\perp$ consists of all values $x = (x_8, \ldots, x_0)$ with $x_0$ and $x_3$ equal to zero. Hence, $\mathbf{x}$ is uniform random on the set of plaintext with these two bits equal to a constant.

Similarly, the vector space $\Lambda_{out}^\perp$ consists of all values $y$ such that the first three bits of $y$ are equal to zero. Hence, the first three bits of $E_k(x)$ are a unique representative for $E_k(x) \bmod \Lambda_{out}^\perp$. ▷

The fact that $F(\mathbf{x}) \bmod \Lambda^\perp$ is uniform random is also called a *saturation* property. Chapter 9 comes back to these properties in the context of saturation attacks.

To compute the false-positive probability of a multidimensional zero-correlation distinguisher, the random permutation model is used as the wrong key randomization hypothesis. The result is the following variant of Corollary 8.2. The proof relies on the probability mass function of the multivariate hypergeometric distribution, which can be derived using an argument analogous to the proof of Theorem 8.1. Exercise 8.5 asks you to give a full proof.

**Theorem 8.3** *Let $\mathbf{F}$ be a uniform random permutation on $\mathbb{F}_2^n$, and let $\Lambda = \Lambda_{\text{in}} \oplus \Lambda_{\text{out}}$ be a multidimensional linear approximation of $\mathbf{F}$ with $|\Lambda| \le 2^n$. The probability that $\Pr_{\mathbf{x}}\left[\mathbf{F}(\mathbf{x}) \equiv t \bmod \Lambda_{\text{out}}^\perp\right] = 1/|\Lambda_{\text{out}}|$ for all $t$ in $\mathbb{F}_2^n$ and with $\mathbf{x}$ uniform random on $s + \Lambda_{\text{in}}^\perp$ for some $s$ in $\mathbb{F}_2^n$ is*

$$\binom{|\Lambda_{\text{out}}^\perp|}{2^n/|\Lambda|}^{|\Lambda_{\text{out}}|} \Big/ \binom{2^n}{|\Lambda_{\text{in}}^\perp|}.$$

## 8.4 Statistical approach

In Section 8.3, it was assumed that correlations or probability distributions (for multidimensional approximations) must be computed exactly to use zero-correlation approximations as distinguishing properties. This requirement can be relaxed, but it comes at the cost of a lower success probability.

Like in ordinary linear cryptanalysis, correlations or – in the multidimensional case – probability distributions can be estimated using a random sample of plaintext-ciphertext pairs. For the correct key, the empirical correlations will be zero on average with a variance of $1/q$. For wrong keys, the average of the empirical correlations is close to but not quite zero, and their variance is approximately $1/q$. These two distributions are the same as in the analysis of (multiple) linear cryptanalysis in the simple model with unknown correlations, except that the roles of right and wrong keys have been reversed.

The average capacity of a multidimensional linear approximation $\Lambda$ of a uniform random permutation $\mathbf{F}$ on $\mathbb{F}_2^n$, is equal to

$$\mathbb{E}_{\mathbf{F}}\, \text{Cap}(\Lambda) = \sum_{\substack{(u,v) \in \Lambda \\ (u,v) \ne (0,0)}} \mathbb{E}_{\mathbf{F}}\left(C_{v,u}^{\mathbf{F}}\right)^2 = |\Lambda|/2^n.$$

Without giving a rigorous proof, we note that $\text{Cap}(\Lambda)$ is close to its mean value with high probability. Hence, in the random permutation model, the capacity of a zero-correlation linear approximation is approximately the same for all wrong keys. By the analysis in Sections 6.1.2 and 7.3.3, the data-complexity for a distinguisher based on a multidimensional zero-correlation linear approximation is

$$q = \sqrt{2|\Lambda|}\, \frac{\Phi^{-1}(P_S) - \Phi^{-1}(P_F)}{\text{Cap}(\Lambda)} \approx \left(\Phi^{-1}(P_S) - \Phi^{-1}(P_F)\right) \frac{2^{n+\frac{1}{2}}}{\sqrt{|\Lambda|}}$$

for $P_S \ge P_F$. The second equality relies on the approximation $\text{Cap}(\Lambda) \approx |\Lambda|/2^n$. It is possible to show that this is optimal in the average-case sense of Section 7.3.1, provided that the random permutation model is used as the wrong key randomization hypothesis.

As explained in Section 6.2.3, if $\Lambda = \Lambda_{\text{in}} \oplus \Lambda_{\text{out}}$, then plaintexts can be sampled from a coset of $\Lambda_{\text{in}}^{\perp}$ to reduce the data-complexity. More precisely, for $P_S \geq P_F$, the data-complexity becomes

$$q = \sqrt{2|\Lambda_{\text{out}}|}\,\frac{\Phi^{-1}(P_S) - \Phi^{-1}(P_F)}{\text{Cap}(\Lambda)} \approx \left(\Phi^{-1}(P_S) - \Phi^{-1}(P_F)\right)\frac{2^{n+\frac{1}{2}}}{|\Lambda_{\text{in}}|\sqrt{|\Lambda_{\text{out}}|}}.$$

Indeed, the average capacity remains the same, but the number of approximations is reduced to $|\Lambda_{\text{out}}|$. Intuitively, the data-complexity is reduced by an additional factor of $\sqrt{|\Lambda_{\text{in}}|}$ on top of the improvement from Section 8.3.2.

## 8.5 Historical remarks

Zero-correlation linear cryptanalysis was introduced by Bogdanov and Rijmen. They used the miss-in-the-middle method, which was discussed in Section 8.2, to find zero-correlation linear approximations.

The statistical approach in Section 8.4, based on multiple zero-correlation linear approximations, is due to Bogdanov and Wang. The chosen-plaintext improvement was first used by Bogdanov, Leander, Nyberg and Wang.

## 8.6 References

Bogdanov, Andrey et al. (Dec. 2012). "Integral and Multidimensional Linear Distinguishers with Correlation Zero." In: *ASIACRYPT 2012*. Ed. by Xiaoyun Wang and Kazue Sako. Vol. 7658. LNCS. Springer, Berlin, Heidelberg, pp. 244–261. DOI: 10.1007/978-3-642-34961-4_16.

Bogdanov, Andrey and Vincent Rijmen (2014). "Linear Hulls with Correlation Zero and Linear Cryptanalysis of Block Ciphers." In: *DCC* 70.3, pp. 369–383. DOI: 10.1007/s10623-012-9697-z.

Bogdanov, Andrey and Meiqin Wang (Mar. 2012). "Zero Correlation Linear Cryptanalysis with Reduced Data Complexity." In: *FSE 2012*. Ed. by Anne Canteaut. Vol. 7549. LNCS. Springer, Berlin, Heidelberg, pp. 29–48. DOI: 10.1007/978-3-642-34047-5_3.

## 8.7 Exercises

### Exercise 8.1

Show that for all $u$ in $\mathbb{F}_2^3$, $(000001000, 000000\|u)$ and $(000001001, 000000\|u)$ are zero-correlation linear approximations for three rounds of the example cipher.

## 8.7 Exercises

### Exercise 8.2

Let $E_k : \mathbb{F}_2^6 \to \mathbb{F}_2^6$ be the construction in Figure 8.2 (see also Figure 2.4). Find a nontrivial zero-correlation linear approximation for all values of $k$.

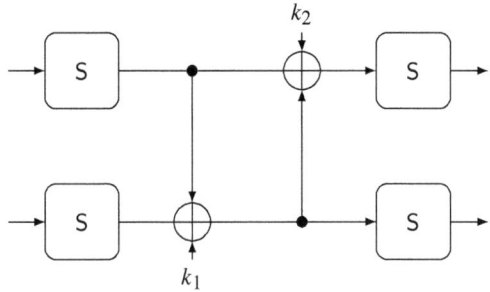

Figure 8.2 A construction with four S-boxes.

### Exercise 8.3

Consider a five round Feistel cipher; the first two rounds are shown in Figure 8.3. Show that if $F_1, \ldots, F_5$ are permutations, then $(0\|u, u\|0)$ is a zero-correlation approximation for all nonzero values of $u$.

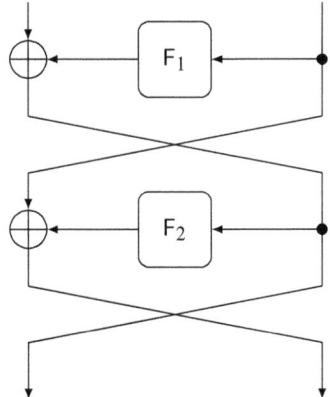

Figure 8.3 Two rounds of a Feistel network.

### Exercise 8.4

The "encrypt-mix-encrypt" construction (Figure 8.4) can be used to build a block cipher based on five functions with only half the block size. Your task is to distinguish the output of the encrypt-mix-encrypt construction from the output of a uniform random permutation on $2n$ bits. In Figure 8.4, $E_1$, $E_2$, $E_3$ and $E_4$ are block ciphers with a block size of $n$ bits and a secret key.

1. Assume that $F\colon \mathbb{F}_2^n \to \mathbb{F}_2^n$ is a permutation. Find a multidimensional zero-correlation approximation of the encrypt-mix-encrypt construction containing $2^{2n}$ linear approximations.
2. Based on your answer above, what are the time and number of chosen plaintexts necessary to distinguish "encrypt-mix-encrypt" from a uniform random permutation on $2n$ bits?

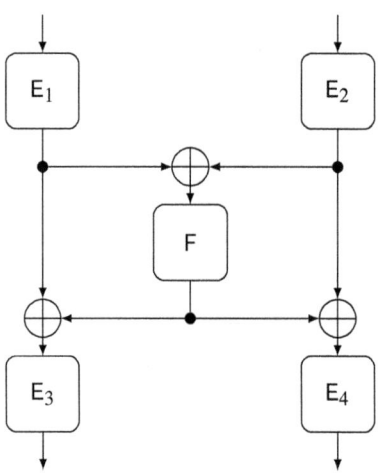

Figure 8.4 The encrypt-mix-encrypt construction.

### Exercise 8.5

Let **F** be a uniform random permutation on $\mathbb{F}_2^n$, and let $\Lambda = \Lambda_{\text{in}} \oplus \Lambda_{\text{out}}$ be a multidimensional linear approximation of **F**.

1. Prove Theorem 8.3.
2. Show that if $\Lambda$ is a multidimensional zero-correlation linear approximation (for an arbitrary function), then $|\Lambda| < 2^n$. Hence, this condition is always satisfied when applying Theorem 8.3 in practice.

### * Exercise 8.6

Corollary 8.2 and Theorem 8.3 are stated for random permutations, but zero-correlation linear cryptanalysis is also applicable to functions that are not invertible.

## 8.7 Exercises

1. Prove analogues of Corollary 8.2 and Theorem 8.3 when **F** is a uniform random function from $\mathbb{F}_2^n$ to $\mathbb{F}_2^m$.
2. Explain why and how the condition $u, v \neq 0$ in Corollary 8.2 can be relaxed when **F** is a uniform random function.

### Exercise 8.7

The goal of this exercise is to find zero-correlation linear attacks on the Rijndael-like example cipher from Section 3.2.1 for a small number of rounds.

1. Find a zero-correlation linear approximation for four rounds.
2. Extend your zero-correlation linear approximation to a multidimensional zero-correlation approximation that requires as little data as possible. What is the resulting data-complexity?
3. Extend your distinguisher to a key-recovery attack on five rounds and estimate the time- and data-complexity.

### * Exercise 8.8

Find a zero-correlation linear approximation for ten rounds of the Rijndael-like example cipher from Section 3.2.1.

### Exercise 8.9

This exercise explores the *key difference invariant correlation* attack. Let $E_k \colon \mathbb{F}_2^n \to \mathbb{F}_2^n$ be a key-alternating block cipher with round keys $k = (k_1, \ldots, k_r)$. That is, $E_k = R_{k_r} \circ \cdots \circ R_{k_2} \circ R_{k_1}$ with $R_{k_i}(x) = R(x) + k_i$.

1. Let $d = d_1 \| \ldots \| d_r \in \mathbb{F}_2^{nr}$ be a difference between two round keys and let
$$\Lambda = \left\{ (u_2, u_3, \ldots, u_{r+1}) \mid u_2, \ldots, u_r \in \mathbb{F}_2^n \text{ and } \prod_{i=1}^r C^R_{u_{i+1}, u_i} \neq 0 \right\}$$
be the set of linear trails with nonzero correlation for the approximation $(u_1, u_{r+1})$. Give sufficient conditions on $d$ and $\Lambda$ to ensure that
$$C^{E_k}_{u_{r+1}, u_1} = C^{E_{k+d}}_{u_{r+1}, u_1},$$
where $k + d = (k_1 + d_1, k_2 + d_2, \ldots, k_r + d_r)$.
2. Give a linear approximation of the example cipher from Section 1.1 with a key-difference invariant correlation, but with a nonzero correlation. Try to maximize the number of rounds.

# 9
# Miscellaneous extensions

The main extensions of linear cryptanalysis were introduced in previous chapters; they are multiple, multidimensional and zero-correlation linear cryptanalysis. However, these are far from the only extensions proposed in the literature. This chapter is a tour of some of the most important proposals.

Most of the extensions of linear cryptanalysis discussed below are partly conjectural: they show how certain combinatorial properties might be used to attack cryptographic primitives, but do not provide a clear way to analyze or find these properties. Chapter 11 returns to this issue.

## 9.1 Exact properties

The correlations of linear approximations are usually only known up to some approximation error, because it is infeasible to take into account all linear trails. As discussed in Chapter 8, zero-correlation linear cryptanalysis is different because it exploits the fact that the correlation of a linear approximation is exactly zero. It turns out that most widely-applicable extensions of linear cryptanalysis (other than multiple and multidimensional) are based on "exact" properties like this.

### 9.1.1 Saturation attacks

As explained in Chapter 8, if $\Lambda = \Lambda_{\text{in}} \oplus \Lambda_{\text{out}}$ with $\Lambda_{\text{in}} \subseteq \mathbb{F}_2^n$ and $\Lambda_{\text{out}} \subseteq \mathbb{F}_2^m$ is a multidimensional zero-correlation linear approximation of $F\colon \mathbb{F}_2^n \to \mathbb{F}_2^m$, then

$$\Pr\left[F(\mathbf{x}) \equiv t \bmod \Lambda_{\text{out}}^{\perp}\right] = \frac{1}{|\Lambda_{\text{out}}|},$$

## 9.1 Exact properties

with **x** uniform random on a coset of $\Lambda_{\text{in}}^{\perp}$ and for all $t$ in $\mathbb{F}_2^m$. In other words, if all elements of a coset of $\Lambda_{\text{in}}^{\perp}$ are encrypted, then every reduction of the ciphertext modulo $\Lambda_{\text{out}}^{\perp}$ occurs an equal number of times. It was already mentioned in Chapter 8 that this is called a saturation property.

Saturation properties can sometimes be found by analyzing the propagation of values rather than linear trails. In fact, this is how the first saturation properties were found before the discovery of zero-correlation linear cryptanalysis. However, such an analysis sometimes also reveals other properties of the ciphertext. For example, some bits of the ciphertext may remain constant when part of the plaintext is saturated. This is illustrated in the following example.

*Example* 9.1  Figure 9.1 propagates a set of plaintexts with one saturated cell through five rounds of the Rijndael-like example cipher from Section 3.2.1. More specifically, the input set consists of eight plaintexts so that all cells except the first are constant, and the first cell takes every value in $\mathbb{F}_2^3$ once. It is not difficult to propagate this set through the first few rounds of the cipher. To do this, we label the cells by "A" if they are equal to every 3-bit value an equal number of times (saturated), "C" if they are constant and "?" otherwise. After five rounds, one of the cells of the state is guaranteed to be constant.

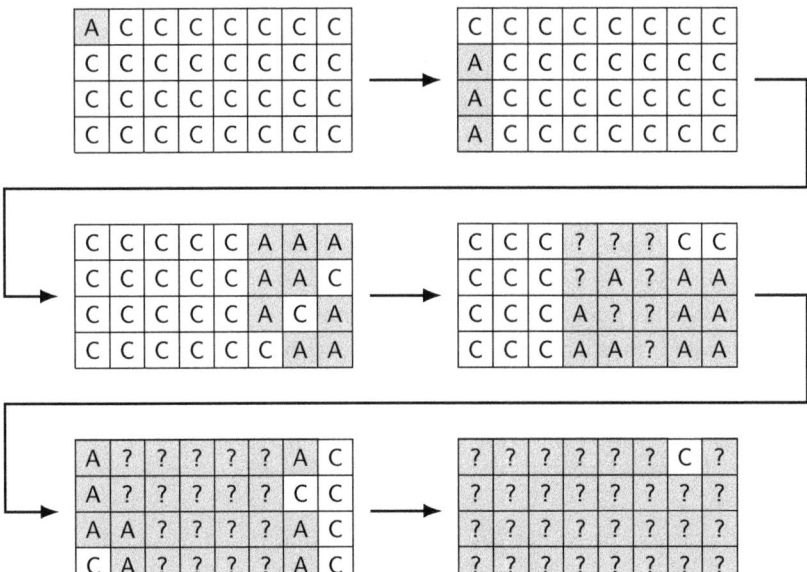

Figure 9.1 Property for five rounds of the Rijndael-like example cipher.

By Corollary 6.6, the above property is equivalent to a multidimensional linear approximation $\Lambda_{\text{in}} \oplus \Lambda_{\text{out}}$. In particular,
$$\Lambda_{\text{in}} = \{x\|00\cdots 0 \mid x \in \mathbb{F}_2^3\}^\perp = \{000\|x \mid x \in \mathbb{F}_2^{93}\}.$$
Furthermore, the set of output masks $\Lambda_{\text{out}}$ consists of all masks that are zero everywhere, except on the constant cell. The multidimensional approximation $\Lambda = \Lambda_{\text{in}} \oplus \Lambda_{\text{out}}$ is not a zero-correlation approximation. Instead, Corollary 6.6 yields
$$\sum_{(u,v)\in\Lambda} (-1)^{u^T s + v^T t} C^F_{v,u} = |\Lambda_{\text{out}}|$$
for all $s$ and a particular $t$ that depends on $s$. Hence, the property in Figure 9.1 corresponds to a large set of linear approximations that sum up to an exceptionally large value. This may seem surprising at first, but the same conclusion can be reached by reasoning about linear trails. ▷

Saturation attacks have been generalized in two different directions. The first direction, called *statistical saturation attacks*, will be discussed in Section 9.2.1. The second direction is called *integral cryptanalysis* and is a research field of itself on par with linear cryptanalysis. It is not discussed in this book.

### 9.1.2 Invariant subspaces

An invariant subspace of a function $\mathsf{F}\colon \mathbb{F}_2^n \to \mathbb{F}_2^n$ is an affine subspace $a + V$ of $\mathbb{F}_2^n$ such that $\mathsf{F}(a+V) \subseteq a+V$. If $\mathsf{F}$ is a permutation, then this means that $\mathsf{F}(a+V) = a + V$. If an invariant subspace exists, it immediately leads to a chosen-plaintext distinguisher with success probability close to one and false-positive probability close to zero. Invariant subspaces can be used for key-recovery attacks by expressing the condition that a partially decrypted ciphertext is an element of $a + V$ as a system of equations. However, this only works if the invariant does not hold for incorrect key guesses.

Historically, invariant subspaces were believed to be a consequence of "obvious" symmetries in the cipher. This is illustrated by following example.

*Example 9.2* Ignoring round constants and keys, there is an invariant subspace for any number of rounds of the Rijndael-like example cipher from Section 3.2.1:

| $x_1$ | $y_1$ | $x_1$ | $y_1$ | $x_1$ | $y_1$ | $x_1$ | $y_1$ |
|---|---|---|---|---|---|---|---|
| $y_1$ | $x_1$ | $y_1$ | $x_1$ | $y_1$ | $x_1$ | $y_1$ | $x_1$ |
| $x_2$ | $y_2$ | $x_2$ | $y_2$ | $x_2$ | $y_2$ | $x_2$ | $y_2$ |
| $y_2$ | $x_2$ | $y_2$ | $x_2$ | $y_2$ | $x_2$ | $y_2$ | $x_2$ |

## 9.1 Exact properties

Here, $x_1$, $x_2$, $y_1$ and $y_2$ are arbitrary values in $\mathbb{F}_2^3$. For most keys, this subspace is not preserved under the round key addition. Nevertheless, there is a set of $2^{12}$ out of $2^{96}$ keys for which this is an invariant subspace. However, the addition of round constants that do not share the same symmetry is guaranteed to prevent the continuation of this property for multiple rounds. ▷

Invariant subspaces that actually depend on the details of the S-box and the linear layer were found for multiple ciphers starting in 2011. These invariant subspaces are usually only valid for a subset of keys, which are then called *weak keys*. The larger the number of weak keys, the greater the success probability of the attack. These "finer" invariant subspaces are typically found using a round-by-round analysis, although not all invariant subspaces can be found in this way.

*Example* 9.3  The Rijndael-like example cipher from Section 3.2.1 has an invariant subspace for $2^{32}$ out of $2^{96}$ keys. In particular, let $U = \{000, 111\}$. Since $S(000) = 111$ and $S(111) = 000$, it also holds that $S(V) = V$. Furthermore,

$$\begin{bmatrix} 0 & I & I & I \\ I & 0 & I & I \\ I & I & 0 & I \\ I & I & I & 0 \end{bmatrix} \begin{bmatrix} 111 \\ 000 \\ 000 \\ 000 \end{bmatrix} = \begin{bmatrix} 000 \\ 111 \\ 111 \\ 111 \end{bmatrix}, \begin{bmatrix} 0 & I & I & I \\ I & 0 & I & I \\ I & I & 0 & I \\ I & I & I & 0 \end{bmatrix} \begin{bmatrix} 000 \\ 000 \\ 111 \\ 111 \end{bmatrix} = \begin{bmatrix} 000 \\ 000 \\ 111 \\ 111 \end{bmatrix},$$

$$\begin{bmatrix} 0 & I & I & I \\ I & 0 & I & I \\ I & I & 0 & I \\ I & I & I & 0 \end{bmatrix} \begin{bmatrix} 000 \\ 111 \\ 111 \\ 111 \end{bmatrix} = \begin{bmatrix} 111 \\ 000 \\ 000 \\ 000 \end{bmatrix}, \begin{bmatrix} 0 & I & I & I \\ I & 0 & I & I \\ I & I & 0 & I \\ I & I & I & 0 \end{bmatrix} \begin{bmatrix} 111 \\ 111 \\ 111 \\ 111 \end{bmatrix} = \begin{bmatrix} 111 \\ 111 \\ 111 \\ 111 \end{bmatrix}.$$

Using the symmetry of the matrix $M$, this implies that $M \bigoplus_{i=1}^{4} U = \bigoplus_{i=1}^{4} U$. Hence, $V = \bigoplus_{i=1}^{32} U$ is an invariant subspace for MixColumns ∘ ShiftRows ∘ SubCells. Since every cell of the rounds constants is equal to 000 or 111, the vector space $V$ is also an invariant subspace of the round constant addition. However, $V$ is only invariant for the key addition step if the round key cells are all equal to 000 or 111. Hence, there are $2^{32}$ weak keys. ▷

The invariant subspace in Example 9.3 was not found in a systematic way. Moreover, even if it would be feasible to list all the invariant subspaces of SubCells, ShiftRows and MixColumns, there could be other invariant subspaces apart from those that SubCells, ShiftRows and MixColumns have in common.

If an invariant subspace is large enough, and if the number of weak keys is not too small, then it is feasible to find it using a black-box approach. Let $F: \mathbb{F}_2^n \to \mathbb{F}_2^n$ be a permutation. To find the smallest invariant subspace of F that contains a value $x$, Algorithm 9.1 can be used. If F has a nontrivial invariant subspace, then repeating Algorithm 9.1 with random choices of $x$ eventually returns this subspace.

---

**Algorithm 9.1** Finding the smallest invariant subspace of a permutation F.

**Input:**
    Permutation $\mathsf{F}\colon \mathbb{F}_2^n \to \mathbb{F}_2^n$
    Vector $x$ in $\mathbb{F}_2^n$

**Output:** The smallest invariant subspace of F containing $x$.

1: ▷ Throughout the algorithm, $U$ can be compactly represented by its basis
2:   $U \leftarrow \{0\}$
3: **repeat**
4:     $V \leftarrow U$
5:     $U \leftarrow \mathrm{Span}\{x + \mathsf{F}(x+z) \mid z \in V\}$
6: **until** $U = V$
7: **return** $x + U$

---

If F has an invariant subspace $a + V$ of dimension $d$, then the probability that a randomly chosen $x$ is in $a + V$ is $2^d/2^n$. Hence, after repeating Algorithm 9.1 $2^{n-d}$ times, it finds (on average) the subspace $a + V$. Throughout Algorithm 9.1, the subspaces $U$ and $V$ can be represented compactly by their basis. Hence, the overall time-complexity is $\mathcal{O}(n^3 \, 2^{n-d})$.

### 9.1.3 Nonlinear invariants

A nonlinear invariant of a function $\mathsf{F}\colon \mathbb{F}_2^n \to \mathbb{F}_2^n$ is a function $f\colon \mathbb{F}_2^n \to \mathbb{F}_2$ such that there exists a constant $b$ in $\mathbb{F}_2$ so that for all $x$ in $\mathbb{F}_2^n$,

$$f(\mathsf{F}(x)) = f(x) + b\,.$$

In other words, $\mathsf{C}(f \circ \mathsf{F}, f) = (-1)^b$. If F is a block cipher, then $b$ may depend on the key.

Every invariant subspace gives rise to a nonlinear invariant with $b = 0$. Indeed, if $a+V$ is an invariant subspace, let $f(x) = 1$ if $x \in a+V$ and $f(x) = 0$ elsewhere. In general, another way to think about a nonlinear invariant $f$ is as a set $S = \{x \in \mathbb{F}_2^n \mid f(x) = 1\}$ such that either $\mathsf{F}(S) \subseteq S$ and $\mathsf{F}(\mathbb{F}_2^n \setminus S) \subseteq \mathbb{F}_2^n \setminus S$ or $\mathsf{F}(S) \subseteq \mathbb{F}_2^n \setminus S$ and $\mathsf{F}(\mathbb{F}_2^n \setminus S) \subseteq S$. The former case corresponds to $b = 0$, the latter to $b = 1$.

The following theorem shows that for certain linear functions represented by a block matrix with identity blocks, it is easy to find nonlinear invariants.

## 9.1 Exact properties

The *degree* of a Boolean function is the degree of its polynomial representation, and a quadratic Boolean function is one of degree two. This definition makes sense because, as you are asked to show in Exercise 9.1, every Boolean function on $\mathbb{F}_2^n$ has a unique polynomial representation in $\mathbb{F}_2[x_1, \ldots, x_n]/(x_1^2 - x_1, \ldots, x_n^2 - x_n)$.

**Theorem 9.1** *Let $M$ be a $bn \times bn$ matrix over $\mathbb{F}_2$ such that $M = A \otimes I$, with $A$ an $n \times n$ matrix over $\mathbb{F}_2$ and $I$ the $b \times b$ identity matrix. If $A$ is an orthogonal matrix, then $f : x_1\| \cdots \|x_n \mapsto \sum_{i=1}^{n} q(x_i)$ is a nonlinear invariant of $x \mapsto Mx$ for every quadratic Boolean function $q$.*

*Proof* If $q$ is a quadratic function, then there exist coefficients $c_{j,k}$ such that $q(z) = \sum_{1 \le j \le k \le b} c_{j,k} z_j z_k$ up to a constant term, which can be assumed to be zero. Note that $q$ can contain linear terms, since $z_j^2 = z_j$. Denote the $j$th bit of $x_i$ in $\mathbb{F}_2^b$ by $x_{i,j}$. At $x = x_1\| \cdots \|x_n$, the function $f$ evaluates to

$$f(x) = \sum_{1 \le j \le k \le b} c_{j,k} \sum_{i=1}^{n} x_{i,j} x_{i,k} = \sum_{1 \le j \le k \le b} c_{j,k} \begin{bmatrix} x_{1,j} & \cdots & x_{n,j} \end{bmatrix} \begin{bmatrix} x_{1,k} \\ \vdots \\ x_{n,k} \end{bmatrix}.$$

Hence, using the fact that $A^{\mathsf{T}} A = I$,

$$f(Mx) = \sum_{1 \le j \le k \le b} c_{j,k} \begin{bmatrix} x_{1,j} & \cdots & x_{n,j} \end{bmatrix} A^{\mathsf{T}} A \begin{bmatrix} x_{1,k} \\ \vdots \\ x_{n,k} \end{bmatrix} = f(x).$$

It follows that $f$ is an invariant of $x \mapsto Mx$. □

Theorem 9.1 leads to a nonlinear invariant for the Rijndael-like example cipher from Section 3.2.1.

*Example 9.4* By Theorem 9.1, every quadratic function of the form $x_1\| \cdots \|x_4 \mapsto \sum_{i=1}^{4} q(x_i)$ is an invariant of the MixColumns matrix $M$ of the Rijndael-like example cipher. The S-box S has the property that $x \mapsto u^{\mathsf{T}} S(x)$ is quadratic for all choices of $u$. Furthermore, for $u = 111$, all $x$ in $\mathbb{F}_2^3$ satisfy

$$u^{\mathsf{T}} S(S(x)) = u^{\mathsf{T}} x.$$

These observations lead to a nonlinear invariant. In particular, let $f$ and $g$ be Boolean functions on $\mathbb{F}_2^{96}$ defined as follows:

$$f(x_1\| \cdots \|x_{32}) = \sum_{i=1}^{32} u^{\mathsf{T}} x = \sum_{i=1}^{32} x_{3i-2} + x_{3i-1} + x_{3i},$$

$$g(x_1\|\cdots\|x_{32}) = \sum_{i=1}^{32} u^\mathsf{T}\mathsf{S}(x) = \sum_{i=1}^{32} x_{3i-2}x_{3i-1} + x_{3i-2}x_{3i} + x_{3i-1}x_{3i}.$$

The round function R satisfies

$$f \circ \mathsf{R} = f \circ \mathsf{MixColumns} \circ \mathsf{ShiftRows} \circ \mathsf{SubCells} = f \circ \mathsf{SubCells} = g.$$

Likewise, the following relation holds over R:

$$g \circ \mathsf{R} = g \circ \mathsf{MixColumns} \circ \mathsf{ShiftRows} \circ \mathsf{SubCells} = g \circ \mathsf{SubCells} = f.$$

Hence, there is a set of weak keys such that either $f(\mathsf{R}_{k_{i+1}}(\mathsf{R}_{k_i}(x))) = f(x)$ for all $x$ in $\mathbb{F}_2^{96}$ or $f(\mathsf{R}_{k_{i+1}}(\mathsf{R}_{k_i}(x))) = f(x) + 1$ for all $x$. In fact, this holds for all choices of $k_{i+1}$. However, every 3-bit cell of $k_i$ must be 000 or 111 for the nonlinear invariant to hold. This follows by a short algebraic manipulation of the polynomial representation of $g$ (see Exercise 9.2). Hence, for $2^{32}$ weak keys, $f$ is a nonlinear invariant for every even number of rounds. ▷

As mentioned in Chapter 2, a large number of linear trails with small absolute correlations can theoretically result in a linear approximation with high absolute correlation. The nonlinear invariant $f$ from Example 9.4 is actually a linear function $x \mapsto u^\mathsf{T}x$. In particular, the fact that either $f(\mathsf{E}_k(x)) = f(x)$ for all $x$ or $f(\mathsf{E}_k(x)) \ne f(x)$ for all $x$ is equivalent to $C_{u,u}^{\mathsf{E}_k} = \pm 1$ for $u = 11\cdots 1$. In Exercise 3.2, it was shown that four-round trails have correlation at most $2^{-16}$. Nevertheless, $(u, u)$ is a linear approximation with correlation $\pm 1$. Although this only holds for a small fraction of $1/2^{64}$ keys, it makes for a good illustration of the limitations of linear trails.

The existence of nonlinear invariants leads to a natural question: how can we analyze the correlation for pairs of Boolean functions, i.e., the nonlinear analogue of linear approximations? Section 9.2.2 discusses some approaches to this problem.

## 9.2 Approximate properties

The downside of the "exact" properties discussed in Section 9.1 is that they are black-or-white: either the property holds, or it does not. In fact, because of this, there is not enough wiggle room for making approximations.

In this section, we discuss some attempts to generalize the properties from Section 9.1 so that they are no longer exact. However, they all suffer from significant difficulties that we will only be able to address in Chapter 11. For this reason, we stick to a high-level description in each case.

## 9.2 Approximate properties

### 9.2.1 Statistical saturation

As explained in Section 9.1.1, saturation attacks rely on encrypting sets of plaintexts so that a part of the plaintext takes all possible values (this part is called "saturated") whereas the remaining part is constant. More generally, the set of plaintexts is a coset of some vector space. In the simplest case, this results in a set of ciphertexts so that part of the output has the saturation property. However, it was shown in Example 9.1 that this can also result in a set of ciphertexts with a constant part.

Statistical saturation attacks exploit the fact that, when a uniform random plaintext from a coset is encrypted, the probability distribution of a part of the ciphertext is nonuniform. The extent of nonuniformity is typically measured by the squared Euclidean imbalance, as this quantity determines the data-complexity. The extreme case, corresponding to the largest squared Euclidean imbalance, is a set of ciphertexts with a constant part.

By Corollary 6.6, statistical saturation properties are equivalent to multidimensional linear approximations. The difference is in how the squared Euclidean imbalance is estimated. In practice, it is usually easier to use linear cryptanalysis for this purpose. However, there are cases where more straightforward arguments based on values are feasible.

### 9.2.2 Nonlinear approximations

A nonlinear approximation of a function $\mathsf{F}\colon \mathbb{F}_2^n \to \mathbb{F}_2^m$ is a pair $(f,g)$ of functions $f\colon \mathbb{F}_2^n \to \mathbb{F}_2$ and $g\colon \mathbb{F}_2^m \to \mathbb{F}_2$. The correlation of $(f,g)$ is defined analogously to the correlation of a linear approximation,

$$\mathsf{C}(g \circ \mathsf{F}, f) = 2\Pr_{\mathbf{x}}\left[g(\mathsf{F}(\mathbf{x})) = f(\mathbf{x})\right] - 1,$$

where the probability is over a uniform random $\mathbf{x}$. If $f = g$ and the correlation is equal to $\pm 1$, then $f$ is a nonlinear invariant of $\mathsf{F}$.

The use of nonlinear approximations was proposed soon after the discovery of linear cryptanalysis, but this did not result in a general way to analyze nonlinear approximations. Numerous other approaches were proposed later on; we mention just one more here. If $\mathsf{F}$ and $\mathsf{G}$ are permutations, then linear approximations of $\mathsf{G} \circ \mathsf{F} \circ \mathsf{G}^{-1}$ are nonlinear approximations of $\mathsf{F}$. Moreover, if $\mathsf{F} = \mathsf{F}_r \circ \cdots \circ \mathsf{F}_2 \circ \mathsf{F}_1$, then a "nonlinear trail" for $\mathsf{F}$ corresponds to a trail for

$$\mathsf{G} \circ \mathsf{F} \circ \mathsf{G}^{-1} = (\mathsf{G} \circ \mathsf{F}_r \circ \mathsf{G}^{-1}) \circ \cdots \circ (\mathsf{G} \circ \mathsf{F}_2 \circ \mathsf{G}^{-1}) \circ (\mathsf{G} \circ \mathsf{F}_1 \circ \mathsf{G}^{-1}).$$

Needless to say, this decomposition is not unique. One approach to nonlinear cryptanalysis is to perform linear cryptanalysis on such an alternative description of the cipher.

All of the aforementioned proposals face significant difficulties, making them unsuitable in practice. It is worth mentioning two recurring issues: (1) the key-dependence of correlations, and (2) there are "too many" nonlinear functions to obtain a useful theory that is on the same level as linear cryptanalysis. Before we can properly discuss these issues, we will need to rebuild the theory of linear cryptanalysis in Chapter 11.

### 9.2.3 Projection framework

It is possible to extend nonlinear approximations to arbitrary functions $f\colon \mathbb{F}_2^n \to X$ and $g\colon \mathbb{F}_2^m \to Y$, with $X$ and $Y$ small sets. These are sometimes called "projection functions."

The general idea of cryptanalysis based on projection functions is to relate $g \circ \mathsf{F}$ and $f$. More concretely, in the known-plaintext setting, one tries to find balanced functions[1] $f$ and $g$ such that $(f(\mathbf{x}), g(\mathsf{F}(\mathbf{x})))$ has a nonuniform distribution for uniform random $\mathbf{x}$. This can again be measured using the squared Euclidean imbalance.

An alternative point of view is that the functions $f$ and $g$ define partitions of $\mathbb{F}_2^n$ and $\mathbb{F}_2^m$. For example, $f$ partitions $\mathbb{F}_2^n$ as

$$\mathbb{F}_2^n = \bigcup_{x \in X} f^{-1}(x),$$

where $f^{-1}(x)$ is the set of values $y$ in $\mathbb{F}_2^n$ such that $f(y) = x$. In *partitioning cryptanalysis*, one studies the relation between a partition of the input space and a partition of the output space. This is equivalent to cryptanalysis based on projection functions.

Partitioning and projection functions make it possible to describe a wide variety of properties. For example, if $f$ and $g$ are linear functions, then they are equivalent to multidimensional linear approximations. However, there are also some notable exceptions, such as multiple linear approximations with a set of masks that does not form a vector space.

Unfortunately, partitioning and projection functions do not do anything more than *describing* properties. They do not, e.g., help us analyze or find these properties.

## 9.3 Historical remarks

Saturation properties were introduced by Knudsen as a part of the "Square attack." Their relation to multidimensional zero-correlation linear approximations was observed by Bogdanov, Leander, Nyberg and Wang.

---

[1] A function is called balanced if every output has the same number of preimages.

Simple examples of invariant subspaces such as Example 9.2 had been observed prior to the discovery of invariant subspaces that depend on the details of the S-box and linear layer by Leander, Abelraheem, AlKhzaimi and Zenner. Algorithm 9.1 is due to Leander, Minaud and Rønjom. Nonlinear invariants and Theorem 9.1 were introduced by Todo, Leander and Sasaki. Example 9.4 is based on Beyne (2018).

Most of the approximate properties discussed in Section 9.2 predate the exact properties from Section 9.1. Statistical saturation attacks were first proposed by Vaudenay; the term was introduced by Collard and Standaert. The use of nonlinear approximations was proposed in 1995 by Harpes, Kramer and Massey under the name "I/O sums," and in 1996 by Knudsen and Robshaw. The approach based on applying linear cryptanalysis to an alternative description of the cipher is due to Beierle, Canteaut and Leander. Partitioning cryptanalysis was proposed by Harpes and Massey, and the notion of projection functions from Section 9.2.3 is due to Wagner. For reasons that will become clear in Chapter 11, none of these proposals really lead to viable extensions of linear cryptanalysis.

## 9.4 References

Beierle, Christof, Anne Canteaut, and Gregor Leander (2018). "Nonlinear Approximations in Cryptanalysis Revisited." In: IACR *Transactions on Symmetric Cryptology* 2018.4, pp. 80–101. ISSN: 2519-173X. DOI: 10.13154/tosc.v2018.i4.80-101.

Beyne, Tim (Dec. 2018). "Block Cipher Invariants as Eigenvectors of Correlation Matrices." In: *ASIACRYPT 2018, Part I*. Ed. by Thomas Peyrin and Steven Galbraith. Vol. 11272. LNCS. Springer, Cham, pp. 3–31. DOI: 10.1007/978-3-030-03326-2_1.

Bogdanov, Andrey et al. (Dec. 2012). "Integral and Multidimensional Linear Distinguishers with Correlation Zero." In: *ASIACRYPT 2012*. Ed. by Xiaoyun Wang and Kazue Sako. Vol. 7658. LNCS. Springer, Berlin, Heidelberg, pp. 244–261. DOI: 10.1007/978-3-642-34961-4_16.

Collard, Baudoin and François-Xavier Standaert (Apr. 2009). "A Statistical Saturation Attack against the Block Cipher PRESENT." In: *CT-RSA 2009*. Ed. by Marc Fischlin. Vol. 5473. LNCS. Springer, Berlin, Heidelberg, pp. 195–210. DOI: 10.1007/978-3-642-00862-7_13.

Daemen, Joan, Lars R. Knudsen, and Vincent Rijmen (Jan. 1997). "The Block Cipher Square." In: *FSE'97*. Ed. by Eli Biham. Vol. 1267. LNCS. Springer, Berlin, Heidelberg, pp. 149–165. DOI: 10.1007/BFb0052343.

Harpes, Carlo, Gerhard G. Kramer, and James L. Massey (May 1995). "A Generalization of Linear Cryptanalysis and the Applicability of Matsui's

Piling-Up Lemma." In: *EUROCRYPT'95*. Ed. by Louis C. Guillou and Jean-Jacques Quisquater. Vol. 921. LNCS. Springer, Berlin, Heidelberg, pp. 24–38. DOI: 10.1007/3-540-49264-X_3.

Harpes, Carlo and James L. Massey (Jan. 1997). "Partitioning Cryptanalysis." In: *FSE'97*. Ed. by Eli Biham. Vol. 1267. LNCS. Springer, Berlin, Heidelberg, pp. 13–27. DOI: 10.1007/BFb0052331.

Knudsen, Lars R. and Matthew J. B. Robshaw (May 1996). "Non-Linear Approximations in Linear Cryptanalysis". In: *EUROCRYPT'96*. Ed. by Ueli M. Maurer. Vol. 1070. LNCS. Springer, Berlin, Heidelberg, pp. 224–236. DOI: 10.1007/3-540-68339-9_20.

Leander, Gregor et al. (Aug. 2011). "A Cryptanalysis of PRINTcipher: The Invariant Subspace Attack." In: *CRYPTO 2011*. Ed. by Phillip Rogaway. Vol. 6841. LNCS. Springer, Berlin, Heidelberg, pp. 206–221. DOI: 10.1007/978-3-642-22792-9_12.

Leander, Gregor, Brice Minaud, and Sondre Rønjom (Apr. 2015). "A Generic Approach to Invariant Subspace Attacks: Cryptanalysis of Robin, iSCREAM and Zorro." In: *EUROCRYPT 2015, Part I*. Ed. by Elisabeth Oswald and Marc Fischlin. Vol. 9056. LNCS. Springer, Berlin, Heidelberg, pp. 254–283. DOI: 10.1007/978-3-662-46800-5_11.

Todo, Yosuke, Gregor Leander, and Yu Sasaki (Dec. 2016). "Nonlinear Invariant Attack – Practical Attack on Full SCREAM, iSCREAM, and Midori64." In: *ASIACRYPT 2016, Part II*. Ed. by Jung Hee Cheon and Tsuyoshi Takagi. Vol. 10032. LNCS. Springer, Berlin, Heidelberg, pp. 3–33. DOI: 10.1007/978-3-662-53890-6_1.

Vaudenay, Serge (Mar. 1996b). "An Experiment on DES Statistical Cryptanalysis." In: *ACM CCS 96*. Ed. by Li Gong and Jacques Stern. ACM Press, New York, pp. 139–147. DOI: 10.1145/238168.238206.

## 9.5 Exercises

### Exercise 9.1

The goal of this exercise is to show that every function $f \colon \mathbb{F}_2^n \to \mathbb{F}_2$ corresponds to a unique polynomial in $\mathbb{F}_2[x_1, \ldots, x_n]/(x_1^2 - x_1, \ldots, x_n^2 - x_n)$. This polynomial is called the *algebraic normal form* of $f$.

1. Show that every polynomial in $\mathbb{F}_2[x_1, \ldots, x_n]$ defines a Boolean function by evaluation.

2. Show that for every Boolean function, there is an interpolating polynomial in $\mathbb{F}_2[x_1, \ldots, x_n]$.
3. Conclude by a counting argument that the "evaluation map," which sends a polynomial to the Boolean function defined by evaluation of the polynomial, is a bijection between $\mathbb{F}_2[x_1, \ldots, x_n]/(x_1^2 - x_1, \ldots, x_n^2 - x_n)$ and the set of all Boolean functions.

### Exercise 9.2

For which keys does the nonlinear invariant from Example 9.4 hold over an arbitrary even number of rounds? Prove your answer.

### Exercise 9.3

A famous Belgian cryptographer often encrypts his personal data using his favorite block cipher. The block cipher has three variants with $r_1 = 10$, $r_2 = 12$ and $r_3 = 14$ rounds. Unfortunately, on this occasion, the cryptographer no longer remembers which of the variants he used.

Fortunately, the cryptographer did ask his students to write down the number of rounds for him. However, in a creative mood, the students decided to encrypt it using a custom cipher $\mathsf{E}_k$ with a 4-bit block size. As illustrated in Figure 9.2, round $i$ of their construction adds the $i$th nibble $k_i$ of the key $k = k_1 \| k_2 \| \cdots \| k_{r+1}$ to the state and then applies the function S given in Table 9.1. Lacking confidence in their own abilities, the students decided to instantiate their cipher $\mathsf{E}_k$ with $r = r_1 \times r_2 \times r_3 + 1 = 1681$ rounds.

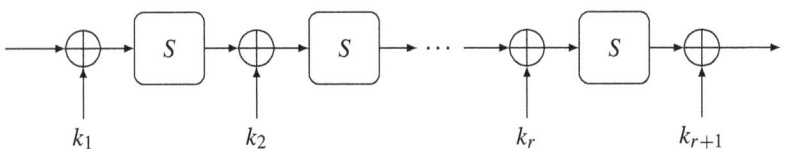

Figure 9.2 The students' encryption method.

Table 9.1. *Lookup table for the function* S

| 0 | 1 | 2 | 3 | 4 | 5 | 6 | 7 | 8 | 9 | a | b | c | d | e | f |
|---|---|---|---|---|---|---|---|---|---|---|---|---|---|---|---|
| 3 | e | 6 | 8 | 0 | c | b | 4 | 1 | d | 5 | a | 7 | 9 | f | 2 |

The students wrote down that the encryption of $r_1 = 10$ is $\mathsf{E}_k(1,0,1,0) = (0,1,0,1)$, i.e., five. They also remember that the ciphertext for $r_2 = 12$ is

$E_k(1,1,0,0) = (0,0,0,0)$. Of course, the students forgot the key, but they still remember that it was an ASCII-encoding of a passphrase consisting only of upper- and lowercase letters. After hearing this, the famous cryptographer exclaims that the students have made a mistake. Can you help the students figure out what is wrong?

# 10
# Functions on Abelian groups

Chapter 11 reconstructs the theory of linear cryptanalysis from a more general point of view. To do this, we need to cover some mathematical ground. We first discuss linear algebra over the field of complex numbers, and then turn to the Fourier analysis of functions on a finite Abelian group. Both of these topics play a central role in Chapter 11.

## 10.1 Linear algebra over $\mathbb{C}$

Linear algebra is concerned with vector spaces and linear transformations between them. However, if the vector spaces are defined over the real or complex numbers, then additional structure enters the picture. This is because the field $\mathbb{C}$ comes with an absolute value function, making it into a metric space.

In this chapter, as well as in Chapter 11, two vector spaces over $\mathbb{C}$ play an important role. They will be used as a running example.

*Example* 10.1   Let $G$ be a finite Abelian group. The free $\mathbb{C}$-vector space on $G$ consists of the set of all formal linear combinations of elements of $G$. That is, every element $u$ of $\mathbb{C}[G]$ is of the form

$$\sum_{x \in G} u_x \, \delta_x,$$

where the values $u_x$ are arbitrary complex numbers and $\delta_x$ is a formal basis vector corresponding to the group element $x$. The basis $\{\delta_x \mid x \in G\}$ is called the standard basis of $\mathbb{C}[G]$.

Similarly, the vector space $\mathbb{C}^G$ consists of all functions $G \to \mathbb{C}$. If $\delta^x$ denotes the function which is one at $x$ and zero everywhere else, then every function $f$ in $\mathbb{C}^G$ can be written as

$$\sum_{x \in G} f(x)\, \delta^x.$$

In Exercise 10.1, you are asked to verify that $\{\delta^x \mid x \in G\}$ is a basis for $\mathbb{C}^G$. It is called the standard basis of $\mathbb{C}^G$. The vector spaces $\mathbb{C}[G]$ and $\mathbb{C}^G$ are both isomorphic to $\mathbb{C}^{|G|}$.  ▷

### 10.1.1 Normed vector spaces and their dual

Vector spaces over $\mathbb{C}$ can be equipped with a norm, which is an abstraction of the idea of "length."

**Definition 10.1** (Normed vector space)  Let $V$ be a vector space over $\mathbb{C}$. A norm on $V$ is a real-valued function $\|\cdot\| : V \to \mathbb{R}$ on $V$ such that

(1) For all $x$ in $V$, $\|x\| \geq 0$ with equality if and only if $x = 0$.
(2) For all $x$ in $V$ and $\lambda$ in $\mathbb{C}$, $\|\lambda x\| = |\lambda|\, \|x\|$.
(3) The triangle-inequality holds: for all $x$ and $y$ in $V$, $\|x + y\| \leq \|x\| + \|y\|$.

A vector space with a norm is called a normed vector space.

*Example* 10.2  For every $p \in [1, \infty)$, the vector space $\mathbb{C}[G]$ can be equipped with the so-called $p$-norm $\|\cdot\|_p$. The $p$-norm of a vector $u$ with coordinates $u_x$ for $x$ in $G$ is defined by

$$\|u\|_p = \sqrt[p]{|G|} \left( \sum_{x \in G} |u_x|^p \right)^{\frac{1}{p}}.$$

A similar norm, also denoted by $\|\cdot\|_p$ with slight abuse of notation, is defined on $\mathbb{C}^G$:

$$\|f\|_p = \frac{1}{\sqrt[p]{|G|}} \left( \sum_{x \in G} |f(x)|^p \right)^{\frac{1}{p}}.$$

For $p = 2$, this is the familiar *Euclidean norm*. Verify that this is indeed a norm. Only the Euclidean norm is used in Chapter 11 – but the general case is helpful to illustrate some ideas. Hence, we omit the proof that the $p$-norm indeed satisfies the properties listed in Definition 10.1 for all $p \geq 1$.  ▷

Every vector space has a dual vector space. Furthermore, the dual of a normed vector space is again a normed vector space. The following definition imposes that $V$ is finite-dimensional to avoid topological subtleties.

## 10.1 Linear algebra over $\mathbb{C}$

**Definition 10.2** (Dual vector space)  Let $V$ be a finite-dimensional vector space over $\mathbb{C}$ with norm $\|\cdot\|$. The dual space $V^\vee$ of $V$ is the $\mathbb{C}$-vector space of all linear functions $V \to \mathbb{C}$, with norm

$$\|f\|^\vee = \max_{\substack{v \in V \\ \|v\| \leq 1}} |f(v)|.$$

The elements of $V^\vee$ are called linear functionals.

Verify that the dual norm defined in Definition 10.2 is indeed a norm. Since $\dim V$ and $\dim V^\vee$ are equal, the vector spaces $V$ and $V^\vee$ are isomorphic. Indeed, one can map a basis of $V$ to a basis of $V^\vee$. The choice of isomorphism is arbitrary because different bases usually result in different isomorphisms. Furthermore, such isomorphisms are in general not isometries, i.e., they do not preserve norms. However, there is a "canonical" isometric isomorphism between $V$ and $V^{\vee\vee}$ that can be specified without such an arbitrary choice of basis.

**Theorem 10.3**  *Let $V$ be a finite-dimensional vector space over $\mathbb{C}$ with norm $\|\cdot\|$. For all $v$ in $V$, define an "evaluation map" $\mathrm{ev}_v \colon V^\vee \to \mathbb{C}$ by $\mathrm{ev}_v(f) = f(v)$. The function $V \to V^{\vee\vee} \colon v \mapsto \mathrm{ev}_v$ is an isomorphism of vector spaces. Furthermore, it is an isometry of normed vector spaces: $\|\mathrm{ev}_v\|^{\vee\vee} = \|v\|$.*

*Proof*  It is not difficult to see that $\mathrm{ev}_{\lambda v} = \lambda \mathrm{ev}_v$ and $\mathrm{ev}_{u+v} = \mathrm{ev}_u + \mathrm{ev}_v$. Hence, $v \mapsto \mathrm{ev}_v$ is a homomorphism of vector spaces. The kernel is zero, and comparing dimensions shows that it must be an isomorphism of vector spaces. To show that it is an isometry, we first prove an upper bound on $\|\mathrm{ev}_v\|^{\vee\vee}$:

$$\|\mathrm{ev}_v\|^{\vee\vee} = \max_{\substack{f \in V^\vee \\ \|f\|^\vee \leq 1}} |f(v)| \leq \|v\|$$

where we have used $|f(v)| \leq \|f\|^\vee \|v\|$. This follows from the definition of $\|f\|^\vee$. Furthermore, there necessarily exists a functional $f$ with $\|f\|^\vee \leq 1$ such that $|f(v)| = \|v\|$. Indeed, let $f(\alpha v) = \alpha \|v\|$ on $\mathrm{Span}\{v\}$ and extend to all of $V$. Such an extension is always possible since $V$ is finite-dimensional, so that $V$ has a basis containing $v$. $\square$

*Example 10.3*  The dual of the vector space $\mathbb{C}[G]$ consists of all linear functions from $\mathbb{C}[G]$ to $\mathbb{C}$. However, every linear function $f \colon \mathbb{C}[G] \to \mathbb{C}$ is determined by its image on the basis vectors $\delta_x$, with $x$ in $G$:

$$f\left(\sum_{x \in G} u_x \delta_x\right) = \sum_{x \in G} u_x f(\delta_x).$$

Hence, linear functions on $\mathbb{C}[G]$ are equivalent to elements of $\mathbb{C}^G$. Up to this canonical isomorphism, $\mathbb{C}^G$ is the same as $\mathbb{C}[G]^\vee$. If $\mathbb{C}[G]$ is equipped with the $p$-norm, then the dual norm on $\mathbb{C}^G$ is the $p/(p-1)$-norm. In Exercise 10.2, you are asked to prove this. The case $p = 2$ is special: the dual of the Euclidean norm is the Euclidean norm. This implies that the map

$$f \mapsto \frac{1}{|G|} \sum_{x \in G} f(x)\delta_x$$

is an isometric isomorphism between $(\mathbb{C}^G, \|\cdot\|_2)$ and $(\mathbb{C}[C], \|\cdot\|_2)$. Since the normed vector space $(\mathbb{C}[G], \|\cdot\|_2)$ is isometrically isomorphic to its dual, it is called self-dual. ▷

### 10.1.2 Inner product spaces

Example 10.3 shows that the Euclidean norm is self-dual. This also follows from the observation that it is induced by an inner product.

**Definition 10.4** (Inner product space)   Let $V$ be a vector space over $\mathbb{C}$. An inner product on $V$ is a function $V \times V \to \mathbb{C}$, denoted by $\langle \cdot, \cdot \rangle$, such that

(1) For all $x$, $y$ and $z$ in $V$ and $\lambda$ and $\mu$ in $\mathbb{C}$, $\langle x, \lambda y + \mu z \rangle = \lambda \langle x, y \rangle + \mu \langle x, z \rangle$.
(2) It is antisymmetric: $\overline{\langle x, y \rangle} = \langle y, x \rangle$ for all $x$ and $y$ in $V$.
(3) For all $x$ in $V$, $\langle x, x \rangle \geq 0$ with equality if and only if $x = 0$.

A vector space with an inner product is called an inner product space.

The next result shows that every inner product space is normed.

**Theorem 10.5**   *If $V$ is a vector space with inner product $\langle \cdot, \cdot \rangle$, then $x \mapsto \|x\| = \sqrt{\langle x, x \rangle}$ is a norm on $V$.*

Moreover, every inner product space is self-dual in the sense of Example 10.3. In the following theorem, an anti-isomorphism of vector spaces over $\mathbb{C}$ is an invertible map $f$ such that $f(\lambda x + \mu y) = \overline{\lambda} f(x) + \overline{\mu} f(y)$ for all vectors $x$ and $y$ and scalars $\lambda$ and $\mu$. The need for anti-isomorphisms is due to the antisymmetry and positivity requirements on inner products.

**Theorem 10.6**   *Let $V$ be a finite-dimensional vector space with inner product $\langle \cdot, \cdot \rangle$. For all $x$ in $V$, define $x^*$ in $V^\vee$ by $x^*(y) = \langle x, y \rangle$ for all $y$ in $V$. The map $x \mapsto x^*$ is an isometric anti-isomorphism.*

*Proof*   See Exercise 10.4. □

*Example 10.4* The standard inner product on $\mathbb{C}^G$ is defined as follows:

$$\langle f, g \rangle = \frac{1}{|G|} \sum_{x \in G} \overline{f(x)} \, g(x).$$

For $G = \mathbb{F}_2^n$, this is the inner product that was used in Exercise 2.1. A similar inner product is defined on $\mathbb{C}[G]$:

$$\langle u, v \rangle = |G| \sum_{x \in G} \overline{u_x} \, v_x,$$

with $u$ and $v$ vectors with coordinates $u_x$ and $v_x$ for $x$ in $G$, respectively.

Due to Theorem 10.3 and 10.6, these inner products lead to isometric anti-isomorphisms between $\mathbb{C}^G$ and $\mathbb{C}[G]$. Hence, if only the structure of the Euclidean norm is considered, then these spaces are equal for all intents and purposes. ▷

Inner product spaces come with a geometric interpretation. Two vectors whose inner product is zero are said to be orthogonal. That is, $u \perp v$ if and only if $\langle u, v \rangle = 0$. A basis consisting of mutually orthogonal vectors with norm one is called an orthonormal basis.

More generally, the modulus of the inner product between two normalized vectors can be interpreted as the cosine of the smallest angle enclosed by them – although for nonreal vectors, some prefer to define the angle as the real part of the inner product. In Exercise 10.7, the concept of angles between vectors is extended to define the angles between two subspaces.

**Theorem 10.7** (Pythagorean theorem) *For all pairs of orthogonal vectors $u$ and $v$ in an inner product space, $\|u + v\|^2 = \|u\|^2 + \|v\|^2$.*

*Proof* The result follows from $\|u + v\|^2 = \langle u + v, u + v \rangle$ and

$$\langle u + v, u + v \rangle = \langle u, u \rangle + \langle u, v \rangle + \langle v, u \rangle + \langle v, v \rangle$$
$$= \|u\|^2 + \|v\|^2 + \langle u, v \rangle + \overline{\langle u, v \rangle}.$$

Since $u$ and $v$ are orthogonal, $\langle u, v \rangle = 0$. □

The orthogonal complement of a subspace $V$ of an inner product space $U$ is the vector space $V^\perp$ of all vectors orthogonal to $V$:

$$V^\perp = \{ u \in U \mid \langle v, u \rangle = 0 \text{ for all } v \in V \}.$$

In Exercise 10.5, you will show that orthogonal complements are also algebraic complements. That is, $U = V \oplus V^\perp$, where $\oplus$ denotes the internal direct sum. Hence, one can define a projection $\pi_V \colon U \to V$ with kernel $V^\perp$. For all $u$ in $U$, $\pi_V(u)$ is the *orthogonal projection* of $u$ on $V$.

The following result, sometimes called the *best approximation theorem*, is an important result related to inner product spaces.

**Theorem 10.8** *Let $V$ be a subspace of a finite-dimensional inner product space $U$. The orthogonal projection of $u$ in $U$ onto $V$ is closest to $u$:*

$$\|u - \pi_V(u)\| = \min_{v \in V} \|u - v\|.$$

*Furthermore, if $u$ is not orthogonal to $V$, then*

$$\frac{|\langle \pi_V(u), u \rangle|}{\|\pi_V(u)\|} = \|\pi_V(u)\| = \max_{\substack{v \in V \\ v \neq 0}} \frac{|\langle v, u \rangle|}{\|v\|}.$$

*Proof* The first claim is that the orthogonal projection minimizes the norm. For all $v$ in $V$, the Pythagorean theorem implies that

$$\|u - v\|^2 = \|u - \pi_V(u) + v - \pi_V(u)\|^2 = \|u - \pi_V(u)\|^2 + \|v - \pi_V(u)\|^2.$$

Hence, if $v \neq \pi_V(u)$, then $\|u - v\| > \|u - \pi_V(u)\|$. The second part of the result follows from the observation that every nonzero $v$ in $V$ satisfies

$$\frac{|\langle u, v \rangle|}{\|v\|} = \frac{|\langle \pi_V(u), v \rangle|}{\|v\|} \leq \|\pi_V(u)\|.$$

This concludes the proof. □

### 10.1.3 Singular value decomposition

The *adjoint* of a linear map $L: U \to V$ between inner product spaces $U$ and $V$ is the linear map $L^\dagger: V \to U$ uniquely defined by the relation

$$\langle L^\dagger(v), u \rangle = \langle v, L(u) \rangle$$

for all $u$ in $U$ and $v$ in $V$. The matrix representation of $L^\dagger$ with respect to two bases is the conjugate-transpose of the matrix representation of $L$ relative to the same bases. This follows from the antisymmetry of inner products.

The linear map $L^\dagger L$ is self-adjoint: $(L^\dagger L)^\dagger = L^\dagger L$. A basic result in linear algebra states that self-adjoint maps are diagonalizable relative to an orthogonal basis (see Exercise 10.8). That is, there exists an orthonormal basis $u_1, \ldots, u_d$ of eigenvectors of $L^\dagger L$. Since $\langle L(u), L(u) \rangle \geq 0$, the corresponding eigenvalues are nonnegative real numbers $\sigma_1^2 \geq \cdots \geq \sigma_d^2$. Let $v_i = L(u_i)/\sigma_i$ for $\sigma_i \neq 0$ and complete to an orthonormal basis $v_1, \ldots, v_d$ of the image of $L$. This shows that every linear map $L: U \to V$ has a *singular value decomposition*.

## 10.1 Linear algebra over $\mathbb{C}$

**Definition 10.9** (Singular value decomposition)  Let $L\colon U \to V$ be a linear map between inner product spaces $U$ and $V$, and let $\sigma_1^2 \geq \cdots \geq \sigma_d^2$ be the eigenvalues of $L^\dagger L$. A singular value decomposition of $L$ consists of orthonormal bases $\{u_1, \ldots, u_d\}$ and $\{v_1, \ldots, v_d\}$ for $U$ and $V$, respectively, such that

$$L(x) = \sum_{i=1}^{d} \sigma_i \langle u_i, x \rangle v_i$$

for all $x$ in $U$. The vectors $u_1, \ldots, u_d$ and $v_1, \ldots, u_d$ are called left and right singular vectors, respectively.

### 10.1.4 Tensor products of vector spaces

A tensor product of $\mathbb{C}$-vector spaces $U$ and $V$ is another $\mathbb{C}$-vector space $U \otimes V$ together with a bilinear map $\otimes\colon U \times V \to U \otimes V$, which has the "universal property" that it uniquely linearizes arbitrary bilinear maps. Specifically, for every $T\colon U \times V \to W$ linear in each variable (bilinear), there exists a unique *linear* map $L\colon U \otimes V \to W$ such that $T(u,v) = L(u \otimes v)$.

This does not uniquely define the tensor product of two vector spaces. However, if two vector spaces $U \otimes_1 V$ and $U \otimes_2 V$ satisfy the universal property mentioned above, then there exists a unique isomorphism $\theta\colon U \otimes_1 V \to U \otimes_2 V$ such that $\otimes_2 = \theta \circ \otimes_1$. This is why we often talk about *the* tensor product.

This is the standard definition of tensor products, but it is abstract. For concrete cases, it is convenient to work with a specific construction of the tensor product. The following example works this out for $\mathbb{C}[G]$ and $\mathbb{C}^G$.

*Example* 10.5  The free vector space $\mathbb{C}[G^2]$ on pairs of elements of $G$ is a tensor product of $\mathbb{C}[G]$ with itself, where the map $\otimes$ is defined by $\delta_x \otimes \delta_y = \delta_{(x,y)}$. Explicitly, $\mathbb{C}[G] \otimes \mathbb{C}[G] = \mathbb{C}[G^2]$.

Similarly, the vector space $\mathbb{C}^{G^2}$ of two-variable functions on $G$ is a tensor product of $\mathbb{C}^G$ with itself, and in this case $\otimes$ is defined by

$$(f \otimes g)(x,y) = f(x)\,g(y).$$

In Section 10.2 and Chapter 11, *the tensor product* will always refer to one of these two specific constructions. ▷

Elements of a tensor product of two vector spaces are sometimes called tensors, and elements of the form $u \otimes v$ are called *elementary* or *rank-one* tensors.

Since the linear functions between vector spaces $U$ and $V$ form a vector space, the tensor product of linear maps is well-defined. The tensor product $L_1 \otimes \cdots \otimes L_n$ of linear maps $L_i : V_i \to U_i$ can be identified canonically with the linear map

$$\bigotimes_{i=1}^{n} V_i \to \bigotimes_{i=1}^{n} U_i$$

$$v_1 \otimes \cdots \otimes v_n \mapsto (L_1 v_1) \otimes \cdots \otimes (L_n v_n).$$

The matrix representation of $L_1 \otimes \cdots \otimes L_n$ relative to bases for $\bigotimes_{i=1}^{n} U_i$ and $\bigotimes_{i=1}^{n} V_i$ that consist of rank-one tensors is the Kronecker product (see Chapter 2) of $n$ matrices.

## 10.2 Fourier analysis on finite Abelian groups

This section focuses on the inner product space $\mathbb{C}^G$ of functions on a finite Abelian group $G$. Given a function $f$ in $\mathbb{C}^G$ and a constant $t$ in $G$, one can define a new function $x \mapsto f(x+t)$ by translation. Another way to phrase this is that the group $G$ acts on $\mathbb{C}^G$. As shown in this section, this action naturally leads to the Fourier transformation.

### 10.2.1 Group characters

The effect of translations on the coordinates of functions in the standard basis of $\mathbb{C}^G$ is inconvenient: the basis vectors are shuffled around by the permutation $\delta^x \mapsto \delta^{x-t}$, which corresponds to multiplication by a permutation matrix. It would be more convenient if the effect of translation would be a simple scaling of the coordinates, i.e., multiplication by a diagonal matrix. This can be achieved by working with respect to a different basis.

To diagonalize the group action, the new basis vectors should be eigenvectors of the set of translation operators. A priori, it is not clear if these operators share a common eigenvector basis. Indeed, it turns out that this is only the case if $G$ is Abelian. A function $\chi : G \to \mathbb{C}$ is a common eigenvector for all translations if and only if $\chi(x+t) = \chi(t) \chi(x)$ for all $x$ and $t$ in $G$. In other words, $\chi$ must be a homomorphism from $G$ to the multiplicative group $\mathbb{C}^\times = \mathbb{C} \setminus \{0\}$. This leads to the following definition.

**Definition 10.10** (Dual group)  Let $G$ be a finite Abelian group. A complex character of $G$ is a group homomorphism $G \to \mathbb{C}^\times$. The Pontryagin dual of $G$ is the group $\widehat{G}$ of all characters of $G$ with respect to the pointwise product.

## 10.2 Fourier analysis on finite Abelian groups

The pointwise product of two characters $\chi$ and $\psi$ is the character $x \mapsto \chi(x)\psi(x)$. Every group has the trivial character $x \mapsto 1$, which acts as a unit for pointwise multiplication. The inverse of $\chi$ is equal to $x \mapsto \chi(-x) = \overline{\chi(x)}$. Hence, $\widehat{G}$ is indeed a group as claimed in Definition 10.10.

The terminology "dual group" is due to the following theorem, which is analogous to Theorem 10.3.

**Theorem 10.11** *Let $G$ be a finite Abelian group. For all $x$ in $G$, define an "evaluation map" $\mathrm{ev}_x \colon \widehat{G} \to \mathbb{C}$ by $\mathrm{ev}_x(\chi) = \chi(x)$. Every evaluation map $\mathrm{ev}_x$ defines a character of $\widehat{G}$. Furthermore, the function $x \mapsto \mathrm{ev}_x$ is an isomorphism of groups from $G$ to the dual of the dual of $G$.*

*Proof* See Exercise 10.9. □

It turns out that $G$ and $\widehat{G}$ are isomorphic, with the caveat that there is no canonical choice of isomorphism. Example 10.6 computes the dual of a cyclic group. This already establishes a special case of the result.

*Example* 10.6 (Dual of a cyclic group)   Let $\mathbb{Z}_n$ denote the additive group of integers modulo $n$. Since $nx = 0$ for all $x$ in $\mathbb{Z}_n$, every character $\chi$ satisfies $\chi(x)^n = 1$. Hence, $\widehat{\mathbb{Z}_n}$ is a group with exponent at most $n$. Furthermore, $\chi(x) = \chi(1)^x$ for all $x$ in $\mathbb{Z}_n$. It follows that $\widehat{\mathbb{Z}_n}$ is a cyclic group of order at most $n$.

Let $\zeta$ be a primitive $n$th root of unity, such as $\zeta = e^{2\pi\sqrt{-1}/n}$. Every function $\chi_u \colon x \mapsto \zeta^{ux}$ is a character of $\mathbb{Z}_n$, because $\chi_u(0) = 1$ and

$$\chi_u(x+y) = \zeta^{u(x+y)} = \zeta^{ux}\zeta^{uy} = \chi_u(x)\chi_u(y).$$

It was shown above that $\widehat{\mathbb{Z}_n}$ has order at most $n$, so the functions $x \mapsto \zeta^{ux}$ are the only characters of $\widehat{\mathbb{Z}_n}$. In fact, $u \mapsto \chi_u$ is an isomorphism between $\mathbb{Z}_n$ and its dual group. The isomorphism depends on the choice of $\zeta$. ▷

The direct sum $G \oplus H$ of groups $G$ and $H$ is a group with underlying set $G \times H$ and group operation $(a,b) + (c,d) = (a+c, b+d)$. The following theorem gives the structure of $\widehat{G \oplus H}$.

**Theorem 10.12** *Let $G$ and $H$ be finite Abelian groups. There is an isomorphism between $\widehat{G \oplus H}$ and $\widehat{G} \oplus \widehat{H}$:*

$$\chi \mapsto (\chi_G, \chi_H),$$

*where $\chi_G$ and $\chi_H$ denote the restriction of $\chi$ to $G$ and $H$, respectively.*

*Proof* Denote the map defined in the theorem by $f$. It is indeed a homomorphism since $f(\chi\psi) = ((\chi\psi)_G, (\chi\psi)_H) = (\chi_G\psi_G, \chi_H\psi_H) = (\chi_G, \chi_H)(\psi_G, \psi_H)$. It is a bijection because its inverse is

$$g\colon (\chi, \psi) \mapsto \big((x, y) \mapsto \chi(x)\psi(y)\big).$$

Indeed, $g(\chi, \psi)_G = \chi$ and $g(\chi, \psi)_H = \psi$. $\square$

The fundamental theorem of finite Abelian groups states that every finite Abelian group is isomorphic to a direct sum of cyclic groups. That is,

$$G \cong \bigoplus_n \mathbb{Z}_n.$$

By Theorem 10.12 and Example 10.6, it follows that

$$\widehat{G} \cong \bigoplus_n \widehat{\mathbb{Z}_n} \cong \bigoplus_n \mathbb{Z}_n \cong G.$$

Hence, every finite Abelian group is isomorphic to its dual. More importantly, by choosing specific isomorphisms and following the chain above in reverse, the elements of the dual group can be found explicitly.

*Example* 10.7  The group $\mathbb{F}_2^n$ is isomorphic to $\mathbb{F}_2 \oplus \mathbb{F}_2 \oplus \cdots \oplus \mathbb{F}_2$. One possible isomorphism is given by $x \mapsto (x_1, x_2, \ldots, x_n)$, where $x_i$ is the $i$th coordinate of $x$ in the standard basis. By Example 10.6, $\widehat{\mathbb{F}_2} = \{\psi_0, \psi_1\}$, where $\psi_u(x) = (-1)^{ux}$. Hence, the inverse of the isomorphism from Theorem 10.12 shows that the characters of $\mathbb{F}_2^n$ are given by

$$\chi_u(x) = \prod_{i=1}^n \psi_{u_i}(x) = (-1)^{u^\mathsf{T} x}.$$

In particular, $u \mapsto \chi_u$ is an isomorphism between $\mathbb{F}_2^n$ and its dual. $\triangleright$

Recall that the original motivation for introducing the dual group was that the characters of a group are the eigenvectors of the translation action of $G$ on $\mathbb{C}^G$. The number of characters of $G$ is equal to $|G|$. Furthermore, the following theorem implies the linear independence of characters. Hence, the characters form a complete eigenvector basis for $\mathbb{C}^G$.

**Theorem 10.13** (Orthogonality of characters)  *For all characters $\chi$ and $\psi$ of a finite Abelian group $G$, it holds that*

$$\langle \chi, \psi \rangle = \begin{cases} 1 & \text{if } \chi = \psi, \\ 0 & \text{otherwise.} \end{cases}$$

*In other words, the characters form an orthonormal basis for $\mathbb{C}^G$.*

## 10.2 Fourier analysis on finite Abelian groups

*Proof* By the definition of the inner product on $\mathbb{C}^G$,

$$\langle \chi, \psi \rangle = \frac{1}{|G|} \sum_{x \in G} \overline{\chi(x)} \psi(x) = \frac{1}{|G|} \sum_{x \in G} (\psi/\chi)(x),$$

where the second equality follows from the fact that $\overline{\chi}$ is the inverse of $\chi$. If $\psi = \chi$, then $\psi/\chi \equiv 1$ and the inner product equals one. If $\psi \neq \chi$, then there exists a value $t$ in $G$ such that $(\psi/\chi)(t) \neq 1$. Hence,

$$\frac{1}{|G|} \sum_{x \in G} (\psi/\chi)(x) = \frac{1}{|G|} \sum_{x \in G} (\psi/\chi)(x+t) = \underbrace{(\psi/\chi)(t)}_{\neq 1} \frac{1}{|G|} \sum_{x \in G} (\psi/\chi)(x).$$

The above equality implies that the inner product is equal to zero. □

### 10.2.2 Fourier transformation

The Fourier transformation is essentially the change-of-basis transformation from the basis of characters to the standard basis. However, in order to avoid choosing an arbitrary isomorphism between $\widehat{G}$ and $G$, it is better to define it as a transformation from $\mathbb{C}^G$ to $\mathbb{C}^{\widehat{G}}$. With this definition, the Fourier transformation maps a character $\chi$ in $\widehat{G} \subset \mathbb{C}^G$ directly to a standard basis vector $\delta^\chi$ in $\mathbb{C}^{\widehat{G}}$. Since group characters are orthogonal (by Theorem 10.13), Definition 10.14 achieves the desired basis transformation.

**Definition 10.14** (Fourier transformation) Let $f: G \to \mathbb{C}$ be a function. The Fourier transformation of $f$ is the function $\widehat{f}: \widehat{G} \to \mathbb{C}$ defined by

$$\widehat{f}(\chi) = \langle \chi, f \rangle = \frac{1}{|G|} \sum_{x \in G} \overline{\chi(x)} f(x).$$

The Fourier transformation is the map $\mathcal{F}: \mathbb{C}^G \to \mathbb{C}^{\widehat{G}}$ defined by $\mathcal{F}(f) = \widehat{f}$.

As discussed above, the Fourier transformation $\mathcal{F}$ from Definition 10.14 maps every character $\chi$ to the corresponding standard basis function $\delta^\chi$. Recall from Example 10.3 that $\mathbb{C}^G$ is dual to $\mathbb{C}[G]$. In particular, the basis of characters has a *dual basis* consisting of the vectors $\chi^*$ defined by

$$\chi^* = \frac{1}{|G|} \sum_{x \in G} \overline{\chi(x)} \delta_x.$$

Note that $\chi^*$ is the result of applying the anti-isomorphism from Theorem 10.6 to $\chi$. By Theorem 10.13, $\psi(\chi^*) = \langle \chi, \psi \rangle = 1$ if $\chi = \psi$ and zero otherwise. This is the property that the term "dual basis" refers to. Accordingly, there exists a dual version of the Fourier transformation, defined by mapping every vector $\chi^*$ to the corresponding standard basis vector $\delta_\chi$. By the orthogonality of characters (Theorem 10.13), the following definition realizes this transformation.

**Definition 10.15** (Fourier transformation, dual)  Let $u$ be a vector in $\mathbb{C}[G]$. The Fourier transformation of $u$ is the vector $\widehat{u}$ in $\mathbb{C}[\widehat{G}]$ defined by

$$\widehat{u}_\chi = \sum_{x \in G} \chi(x) u_x.$$

The Fourier transformation is the map $\mathcal{F} \colon \mathbb{C}[G] \to \mathbb{C}[\widehat{G}]$ defined by $\mathcal{F}(u) = \widehat{u}$.

The "dual" Fourier transformation $\mathcal{F}$ from Definition 10.15 is related to the Fourier transformation $\mathcal{F}$ from Definition 10.14 by $\mathcal{F} = \mathcal{F}^{-\vee}$. Here, $\mathcal{F}^\vee \colon \mathbb{C}[\widehat{G}] \to \mathbb{C}[G]$ is the transpose of $\mathcal{F}$, which is defined by the relation

$$f\big(\mathcal{F}^\vee(u)\big) = \big(\mathcal{F}(f)\big)(u)$$

for all $u$ in $\mathbb{C}[\widehat{G}]$ and $f$ in $\mathbb{C}^G$. Setting $u = \mathcal{F}^{-\vee}(\chi^*)$ and $f = \psi$ equal to a character of $G$, this yields $\delta^\psi(u) = \psi(\chi^*)$. Hence, $u = \delta_\chi$ and $\mathcal{F} = \mathcal{F}^{-\vee}$.

*Example* 10.8  The Fourier transformation from Definition 5.2 is the Fourier transformation on $\mathbb{C}[\mathbb{F}_2^n]$. Indeed, by Example 10.7, the characters of $\mathbb{F}_2^n$ are

$$\chi_u(x) = (-1)^{u^\mathsf{T} x}.$$

Plugging this into Definition 10.15 yields Definition 5.2. ▷

The vector space $\mathbb{C}^{\widehat{G}}$ is an inner product space with

$$\langle f, g \rangle = \sum_{\chi \in \widehat{G}} \overline{f(\chi)} g(\chi)$$

for all $f$ and $g$ in $\mathbb{C}^{\widehat{G}}$. With this choice of inner product, the Fourier transformation is unitary. A similar result holds for the transformation $\mathcal{F}$.

**Theorem 10.16**  *The Fourier transformation* $\mathcal{F} \colon \mathbb{C}^G \to \mathbb{C}^{\widehat{G}}$ *is unitary. That is, $\mathcal{F}^{-1} = \mathcal{F}^\dagger$, with $\mathcal{F}^\dagger$ the adjoint of $\mathcal{F}$. Explicitly, if $\widehat{f} = \mathcal{F}(f)$, then*

$$f(x) = \sum_{\chi \in \widehat{G}} \chi(x) \widehat{f}(\chi).$$

*Proof*  By definition, the adjoint of $\mathcal{F}$ satisfies $\langle \mathcal{F}^\dagger(g), f \rangle = \langle g, \mathcal{F}(f) \rangle$ for all $g$ in $\mathbb{C}^{\widehat{G}}$ and $f$ in $\mathbb{C}^G$. Hence, for all $\chi$ and $\psi$,

$$\langle (\mathcal{F}^\dagger \mathcal{F})(\chi), \psi \rangle = \langle \mathcal{F}(\chi), \mathcal{F}(\psi) \rangle = \langle \delta^\chi, \delta^\psi \rangle = \delta^\chi(\psi).$$

It follows that $\mathcal{F}^{-1} = \mathcal{F}^\dagger$. For the concrete formula, note that

$$f(x) = \big(\mathcal{F}^{-1} \widehat{f}\big)(\delta_x) = \widehat{f}\big(\mathcal{F} \delta_x\big) = \sum_{\chi \in \widehat{G}} \chi(x) \widehat{f}(\chi).$$

The second equality follows from Definition 10.15.  □

## 10.2.3 Pontryagin duality

The relation between $G$ and $\widehat{G}$, or between $\mathbb{C}^G$ and $\mathbb{C}^{\widehat{G}}$, is called *Pontryagin duality*. This duality carries over to the subgroups of $G$ and $\widehat{G}$.

**Definition 10.17** (Annihilator)  Let $G$ be a finite Abelian group. The annihilator of a subset $H$ of $G$ is the subgroup

$$H^\perp = \{\chi \in \widehat{G} \mid \forall x \in H : \chi(x) = 1\}.$$

Similarly, for a subgroup $H$ of $\widehat{G}$, the annihilator of $H$ is the subgroup

$$H^\perp = \{x \in G \mid \forall \chi \in H : \chi(x) = 1\}.$$

These definitions are equivalent up to the canonical isomorphism from Theorem 10.11.

It follows from Definition 10.17 that $H^{\perp\perp} = H$.

If $\{0\} \subseteq H \subseteq K \subseteq G$, then $\{1\} \subseteq K^\perp \subseteq H^\perp \subseteq \widehat{G}$. In other words, "taking annihilators" maps subgroups of $G$ to subgroups of $\widehat{G}$ and conversely, but reverses the order of inclusion. The following result gives a more detailed characterization of the groups $H^\perp$ and $K^\perp$.

**Theorem 10.18**  *Let $H$ be a subgroup of $G$. There is an isomorphism between $G/H$ and $\widehat{H^\perp}$, given by sending $x + H$ to the evaluation map $\mathrm{ev}_x : H^\perp \to \mathbb{C}$. Furthermore, this isomorphism leads to the following equality of subspaces:*

$$\mathrm{Span}\{\chi \mid \chi \in H^\perp\} = \mathrm{Span}\{f \circ \pi_H \mid f \in \mathbb{C}^{G/H}\},$$

*where $\pi_H : G \to G/H$ is the projection map $\pi_H(x) = x + H$.*

*Proof*  Observe that $\theta : G \to \widehat{H^\perp}$ defined by $\theta(x) = \mathrm{ev}_x$ is a homomorphism of groups. By Theorem 10.11, $\mathrm{ev}_x$ is the character of $H^\perp$ defined by $\chi \mapsto \chi(x)$. Hence, $\mathrm{ev}_x \equiv 1$ if and only if $x$ is an element of $H$. It follows that the kernel of $\theta$ is $H$. The first isomorphism theorem for groups then shows that $x + H \mapsto \mathrm{ev}_x$ is an isomorphism from $G/H$ to $\widehat{H^\perp}$.

The equality between the spaces can be demonstrated as follows. Every function in the span of $H^\perp$ is constant on the cosets of $H$, because $\chi(x) = \chi(y)$ if and only if $x/y \in H^\perp$. Hence,

$$\mathrm{Span}\{\chi \mid \chi \in H^\perp\} \subseteq \mathrm{Span}\{f \circ \pi_H \mid f \in \mathbb{C}^{G/H}\}.$$

However, by the isomorphism, the dimensions match, so this is an equality. □

Theorem 10.18 simplifies the calculation of the Fourier transformation of functions that are constant on the cosets of a subgroup $H$ of $G$, i.e., of the form $f \circ \pi_H$ with $f$ in $\mathbb{C}^{G/H}$. In Fourier analysis, such functions are called *periodic*. By Theorem 10.18, $f \circ \pi_H$ is a linear combination of the characters in $H^1$. Hence,

$$\widehat{f \circ \pi_H}(\chi) = \begin{cases} \widehat{f}(\chi_{G/H}) & \text{if } \chi \in H^1, \\ 0 & \text{else.} \end{cases}$$

In the first case above, $\chi_{G/H}$ is the character of $G/H$ obtained from $\chi$ by $x + H \mapsto \chi(x)$. This is well defined because $\chi \in H^1$. Indeed, by Definition 10.14,

$$\widehat{f \circ \pi_H}(\chi) = \frac{1}{|G|} \sum_{x \in G} \overline{\chi(x)} f(\pi_H(x))$$

$$= \frac{1}{|G/H|} \sum_{x+H \in G/H} \overline{\chi(x+H)} f(x+H)$$

$$= \widehat{f}(\chi_{G/H}).$$

The equality of subspaces in Theorem 10.18 can be dualized by applying the anti-isomorphism $x \mapsto x^*$ from Theorem 10.6 to both sides:

$$\mathsf{Span}\{\chi^* \mid \chi \in H^1\} = \mathsf{Span}\{\sum_{g \in x+H} \delta_g \mid x + H \in G/H\}.$$

There is a variant of Theorem 10.18 for subgroups of $\widehat{G}$; see Exercise 10.10.

## 10.3 Historical remarks

Additional details about the linear algebra discussed in this chapter may be found in most linear algebra textbooks, such as Halmos (1958). The theory of Fourier analysis on finite Abelian groups is developed in multiple books, including Terras (1999).

The linear algebra introduced in this chapter is necessary for Chapter 11. The motivation to study the Fourier transformation for arbitrary Abelian groups, and not just for $\mathbb{F}_2^n$, is that Chapter 11 reconstructs the theory of linear cryptanalysis in this setting – although this is by no means the point of the chapter. Historically, linear cryptanalysis was first extended to other finite Abelian groups by Baignères, Stern and Vaudenay. A more complete treatment was given in Beyne (2021).

## 10.4 References

Baignères, Thomas, Jacques Stern, and Serge Vaudenay (Aug. 2007). "Linear Cryptanalysis of Non Binary Ciphers." In: *SAC 2007*. Ed. by Carlisle M.

Adams, Ali Miri, and Michael J. Wiener. Vol. 4876. LNCS. Springer, Berlin, Heidelberg, pp. 184–211. DOI: 10.1007/978-3-540-77360-3_13.

Beyne, Tim (Dec. 2021). "A Geometric Approach to Linear Cryptanalysis." In: *ASIACRYPT 2021, Part I*. Ed. by Mehdi Tibouchi and Huaxiong Wang. Vol. 13090. LNCS. Springer, Cham, pp. 36–66. DOI: 10.1007/978-3-030-92062-3_2.

Halmos, Paul R. (1958). *Finite-dimensional Vector Spaces*. 1st ed. Undergraduate Texts in Mathematics. Springer New York, NY.

Terras, Audrey (1999). *Fourier Analysis on Finite Groups and Applications*. London Mathematical Society Student Texts. Cambridge University Press, Cambridge.

## 10.5 Exercises

### Exercise 10.1

Prove that the functions $\delta^x$ with $x$ in $G$ form a basis for $\mathbb{C}^G$:
1. Show that $\mathrm{Span}\{\delta^x \mid x \in G\} = \mathbb{C}^G$.
2. Show that the functions $\delta^x$ with $x$ in $G$ are linearly independent.

### Exercise 10.2

Let $p$ and $q$ be real numbers greater than one such that $1/p + 1/q = 1$.
1. Show that $xy \leq x^p/p + y^q/q$ for all nonnegative $x$ and $y$ in $\mathbb{R}$.
2. Deduce that $|f(g)| \leq \|f\|_q \|g\|_p$ for all $f$ in $\mathbb{C}^G$ and $g$ in $\mathbb{C}[G]$.
3. Show that $\|f\|_p^\vee \leq \|f\|_q$.
4. For all $f$ in $\mathbb{C}^G$, construct a $g$ so that $|f(g)| = \|f\|_q \|g\|_p$. Based on this, conclude that $\|f\|_p^\vee = \|f\|_q$.

### Exercise 10.3

Prove that $\langle f, g \rangle = \frac{1}{|G|} \sum_{x \in G} \overline{f(x)} g(x)$ defines an inner product on $\mathbb{C}^G$.

### Exercise 10.4

The goal of this exercise is to prove Theorem 10.6. Let $\theta : x \mapsto x^*$.
1. Prove that $\theta(x + y) = \theta(x) + \theta(y)$.
2. Prove that $\theta(\lambda x) = \overline{\lambda} \theta(x)$.
3. Prove that $\theta$ is invertible. What is $\theta^{-1}$?

### Exercise 10.5

Let $V$ be a subspace of a finite-dimensional inner product space $U$. Prove that:

1. $V^\perp$ is an inner product space.
2. The orthogonal complement of $V^\perp$ is $V$ itself: $V^{\perp\perp} = V$.
3. $U = V \oplus V^\perp$.

The last result shows that orthogonal complements are algebraic complements.

### Exercise 10.6

Let $V$ be a subspace of a finite-dimensional inner product space $U$. The annihilator of $V$ is the subspace $V^0$ of $U^\vee$ defined by

$$V^0 = \{u \in U^* \mid \forall v \in V: u(v) = 0\}.$$

Dually, the annihilator of a subspace $W$ of $U^\vee$ is the following subspace of $U$:

$$W^0 = \{u \in U \mid \forall w \in W: w(u) = 0\}.$$

Prove that:

1. $\dim V + \dim V^0 = \dim U = \dim U^* = \dim W + \dim W^0$.
2. If $U$ is an inner product space, then $(V^0)^* = V^\perp$ and $(W^0)^* = W^\perp$.

For the second question, $x \mapsto x^*$ is the anti-isomorphism from Theorem 10.6.

### Exercise 10.7

The concept of angles between vectors generalizes to subspaces of a finite-dimensional inner product space $W$. For subspaces $U$ and $V$ of $W$, define a linear map $\langle V, U \rangle : U \to V$ by $\langle V, U \rangle = \pi_V \iota_U$, where $\iota_U : U \to W$ is the inclusion map and $\pi_V : W \to V$ is the orthogonal projection on $V$.

1. Show that if $U$ and $V$ are one-dimensional subspaces spanned by unit-norm vectors $u$ and $v$, respectively, then $\langle V, U \rangle : \lambda u \mapsto \langle v, u \rangle \lambda v$.
2. Show that for all vectors $u$ in $U$, no other vector in $V$ of the same length makes a smaller angle to $u$ than $\langle V, U \rangle (u)$.
3. Let $\sigma_1, \ldots, \sigma_d$ be the singular values of $\langle V, U \rangle$, corresponding to right and left singular vectors $u_1, \ldots, u_d$ and $v_1, \ldots, v_d$, respectively. Let $U_i = U \cap \mathrm{Span}\{u_1, \ldots, u_{i-1}\}^\perp$ and $V_i = V \cap \mathrm{Span}\{v_1, \ldots, v_{i-1}\}^\perp$. Prove that for all $i$ in $\{1, \ldots, d\}$,

$$\sigma_i = \frac{\langle u_i, v_i \rangle}{\|u_i\| \, \|v_i\|} = \max_{\substack{u \in U_i \setminus \{0\} \\ v \in V_i \setminus \{0\}}} \frac{|\langle u, v \rangle|}{\|u\| \, \|v\|}.$$

The angles $0 \leq \theta_1 \leq \cdots \leq \theta_d \leq \pi/2$ such that $\cos\theta_i = \sigma_i$ are called the *principal angles* between the subspaces $U$ and $V$. The singular vectors are the directions along which the principal angles are measured.

### * Exercise 10.8

Let $L: V \to V$ be a linear map on a finite-dimensional inner product space $V$. A subspace $U$ of $V$ is called an invariant subspace of $L$ if $L(U) \subseteq U$.

1. Prove that if $U$ is an invariant subspace of $L$, then $U^\perp$ is an invariant subspace of $L^\dagger$ and conversely.
2. Use the previous result to prove that if $L$ is self-adjoint, then there exists an orthonormal basis of $V$, so that the matrix representation of $L$ relative to this basis is diagonal.

### Exercise 10.9

Prove that $\widehat{\widehat{G}}$ is canonically isomorphic to $G$ through the isomorphism $x \mapsto \text{ev}_x$, where $\text{ev}_x: \chi \mapsto \chi(g)$ is the evaluation map (see Theorem 10.11).

1. Show that $\text{ev}_x$ is a character of $\widehat{G}$ for all $x$ in $G$.
2. Prove that $x \mapsto \text{ev}_x$ is an isomorphism of groups.

### Exercise 10.10

The goal of this exercise is to prove an analogue of Theorem 10.18 for a subgroup $H$ of $\widehat{G}$. For each question, use two different arguments: one similar to the proof of Theorem 10.18, and the other based on duality.

1. Give an isomorphism from $\widehat{G}/H$ to $\widehat{H^1}$.
2. Show that this isomorphism lifts to the following equality of subspaces:

$$\text{Span}\{\delta_x \mid x \in H^1\} = \text{Span}\{\sum_{\psi \in \chi H} \psi^* \mid \chi H \in \widehat{G}/H\}.$$

# 11
# Geometric approach

In this chapter, we rebuild the theory of linear cryptanalysis one last time. One of the reasons for doing this was already mentioned in Chapter 9: there are various combinatorial properties that might be useful, but for which there are no analytic methods. However, before attempting to address this issue, we must take a step back and try to improve our understanding of linear cryptanalysis.

## 11.1 Geometric viewpoint

Let $F: G \to H$ be a cryptographic primitive, such as a block cipher. The starting point of the geometric approach is a reformulation of cryptanalytic properties of $F$, such as linear approximations, in terms of pairs of vector spaces of $\mathbb{C}[G]$ and $\mathbb{C}^H$. This viewpoint is quite general and, with some adaptations, also applies to other important methods such as differential and integral cryptanalysis.

### 11.1.1 Cryptanalytic properties

In the simplest case, the vector spaces are one-dimensional, so that the cryptanalytic property is determined by a pair $(u, v)$, where $u$ is an element of $\mathbb{C}[G]$ and $v$ an element of $\mathbb{C}^H$. Intuitively, $u$ and $v$ have the following meaning:

- The vector $u$ represents an assignment of weights (complex numbers) to the elements of $G$. This is a way to keep track of the state of a collection of inputs or outputs.
- The function $v$ maps elements of $G$, and by extension $\mathbb{C}[G]$, to $\mathbb{C}$. It represents a measurement or observation of the state of a collection of inputs or outputs.

## 11.1 Geometric viewpoint

Applying a function $\mathsf{F}\colon G \to H$ to the state transforms the assignment of weights on $G$ to a corresponding assignment on $H$. In the simplest case, when F is a permutation, it leads to a rearrangement of the weights that were assigned to elements of $X$. Section 11.1.2 describes the effect of general functions F. For now, it is sufficient to say that the result is characterized by a linear function $T^\mathsf{F}\colon \mathbb{C}[G] \to \mathbb{C}[H]$ that maps $\delta_x$ to $\delta_{\mathsf{F}(x)}$.

In cryptanalysis, it is rarely possible to compute the exact state $T^\mathsf{F} u$. However, it is enough to evaluate $v(T^\mathsf{F} u)$. The following definition generalizes the above.

**Definition 11.1** (Cryptanalytic property)  A cryptanalytic property of a function $\mathsf{F}\colon G \to H$ is a pair $(U,V)$ with $U$ a subspace of $\mathbb{C}[G]$ and $V$ a subspace of $\mathbb{C}^H$. The evaluation of the property at $u$ in $U$ and $v$ in $V$ is equal to $v(T^\mathsf{F} u)$.

The purpose of the techniques introduced in this chapter is to estimate $v(T^\mathsf{F} u)$. The following example of a cryptanalytic property is useful to keep in mind.

*Example* 11.1   Let $\mathsf{F}\colon G \to H$ be a function, and $X$ and $Y$ subsets of $G$ and $H$, respectively. This example defines a cryptanalytic property that evaluates to the number of $x$ in $X$ such that $\mathsf{F}(x) \in Y$. Let $\delta_X$ be the vector defined by

$$\delta_X = \sum_{x \in X} \delta_x.$$

A concrete example for $G = \mathbb{F}_2^2$ is shown in Figure 11.1. Define $\delta^Y\colon H \to \mathbb{C}$ by

$$\delta^Y = \sum_{y \in Y} \delta^y.$$

That is, $\delta^Y(y) = 1$ if $y \in Y$ and $\delta^Y(y) = 0$ otherwise. Extending $\delta^Y$ linearly to $H$ gives a linear functional that sums its input over $X$. Hence, the property $(U,V)$ with $U = \mathrm{Span}\{\delta_X\}$ and $V = \mathrm{Span}\{\delta^Y\}$ evaluates to

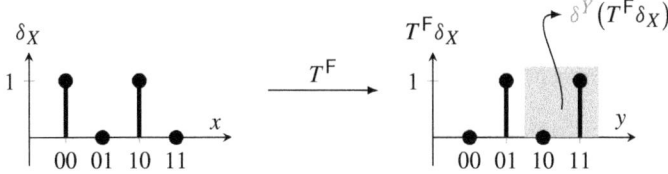

Figure 11.1  Property of $\mathsf{F}\colon \mathbb{F}_2^2 \to \mathbb{F}_2^2$ from Example 11.1 with $\mathsf{F}(x) = x + 11$, $X = \{00, 10\}$ and $Y = \{10, 11\}$. The evaluation is $\delta^Y(T^\mathsf{F} \delta_X) = 1$.

$$\delta^Y(T^F \delta_X) = \sum_{x \in X} \delta^Y(\delta_{F(x)}) = |\{x \in X \mid F(x) \in Y\}|.$$

The first equality uses the fact that $T^F$ maps $\delta_x$ to $\delta_{F(x)}$.  ▷

### 11.1.2 Propagation

For a general function $F: G \to H$, the linear map $T^F$ is defined as follows.

**Definition 11.2** (Pushforward)  Let $F: G \to H$ be a function. The pushforward operator of F is the linear map $T^F: \mathbb{C}[G] \to \mathbb{C}[H]$ defined by

$$T^F \delta_x = \delta_{F(x)}$$

for all $x$ in $G$.

Definition 11.2 implies that for $u$ in $\mathbb{C}[G]$ with coordinates $u_x$,

$$T^F u = \sum_{y \in H} \delta_y \sum_{\substack{x \in G \\ F(x) = y}} u_x.$$

The matrix representation of $T^F$ with respect to the standard bases of $\mathbb{C}[G]$ and $\mathbb{C}[H]$ is called the *transition matrix* of F. With some abuse of notation, it is denoted by $T^F$ as well. The coordinates of the transition matrix are given by

$$T^F_{y,x} = \begin{cases} 1 & \text{if } y = F(x), \\ 0 & \text{else.} \end{cases}$$

There is a dual version of Definition 11.2, corresponding to the transpose of $T^F$. It describes the backward propagation of a function $v$ in $\mathbb{C}^H$ through F.

**Definition 11.3** (Pullback)  Let $F: G \to H$ be a function. The pullback operator of F is the linear map $T^{F^\vee}: \mathbb{C}^H \to \mathbb{C}^G$ defined by

$$T^{F^\vee} \delta^y = \delta^y \circ F$$

for all $y$ in $H$.

The pullback operator is indeed the transpose of the pushforward operator:

$$(T^{F^\vee} \delta^y)(\delta_x) = (\delta^y \circ F)(x) = \delta^y(\delta_{F(x)}) = \delta^y(T^F \delta_x).$$

This also means that the matrix representation of $T^{F^\vee}$ with respect to the standard bases of $\mathbb{C}^H$ and $\mathbb{C}^G$ is the transpose of the transition matrix of F.

## 11.1 Geometric viewpoint

*Example* 11.2  The transition matrix of the S-box of the example cipher from Chapter 1 is equal to the following permutation matrix:

$$T^S = \begin{bmatrix} 0 & 0 & 0 & 0 & 0 & 0 & 0 & 1 \\ 0 & 0 & 0 & 0 & 1 & 0 & 0 & 0 \\ 0 & 1 & 0 & 0 & 0 & 0 & 0 & 0 \\ 0 & 0 & 0 & 0 & 0 & 0 & 1 & 0 \\ 0 & 0 & 1 & 0 & 0 & 0 & 0 & 0 \\ 0 & 0 & 0 & 1 & 0 & 0 & 0 & 0 \\ 0 & 0 & 0 & 0 & 0 & 1 & 0 & 0 \\ 1 & 0 & 0 & 0 & 0 & 0 & 0 & 0 \end{bmatrix}.$$

The matrix representation of the pullback operator is the transpose of $T^S$. ▷

The following theorem summarizes two important properties of pushforward operators. All of these properties also hold for pullback operators, but for (2) the order of multiplication should be reversed.

**Theorem 11.4**  *Let* $F: G \to H$ *be a function. The pushforward operator* $T^F$ *of* F *has the following properties:*

(1) *If* $F((x_1, \ldots, x_n)) = (F_1(x_1), \ldots, F_n(x_n))$ *with* $F_i: G_i \to H_i$ *so that* $G = \bigoplus_{i=1}^n G_i$ *and* $H = \bigoplus_{i=1}^n H_i$, *then* $T^F = T^{F_1} \otimes T^{F_2} \otimes \cdots \otimes T^{F_n}$.
(2) *If* $F = F_r \circ \cdots \circ F_2 \circ F_1$, *then* $T^F = T^{F_r} \cdots T^{F_2} T^{F_1}$.

*Proof*  It is sufficient to prove the result for $n = 2$ and $r = 2$. It follows from Definition 11.2 and the definition of tensor products in Example 10.5 that

$$T^F \underbrace{\delta_{(x_1, x_2)}}_{\delta_{x_1} \otimes \delta_{x_2}} = \underbrace{\delta_{(F_1(x_1), F_2(x_2))}}_{\delta_{F_1(x_1)} \otimes \delta_{F_2(x_2)}}.$$

Hence, for all $x_1$ in $G_1$ and $x_2$ in $G_2$,

$$T^F(\delta_{x_1} \otimes \delta_{x_2}) = T^{F_1}\delta_{x_1} \otimes T^{F_2}\delta_{x_2}.$$

The second property is also an immediate consequence of Definition 11.2:

$$T^{F_2 \circ F_1}\delta_x = \delta_{F_2(F_1(x))} = T^{F_2}\delta_{F_1(x)} = T^{F_2}T^{F_1}\delta_x,$$

and this holds for all $x$ in $G$. □

### 11.1.3 Geometry

In Chapter 10, the vector spaces $\mathbb{C}[G]$ and $\mathbb{C}^G$ were equipped with the $p$-norm. As discussed, the special case $p = 2$ leads to an inner product space. It turns out that the Euclidean norm plays an important role in cryptanalysis.

If $(x_1, y_1), \ldots, (x_q, y_q)$ are $q$ plaintext-ciphertext samples, then an unbiased estimator of $v(T^F u)$ is given by

$$t = \frac{|G|}{q} \sum_{i=1}^{q} v(y_i) u_{x_i}.$$

Assume that the inputs are sampled independently and uniformly at random. A variant of the simple model postulates that for wrong keys, the outputs $\mathbf{y}_1, \ldots, \mathbf{y}_q$ are independent and uniform random variables on $H$. If, in addition, $\sum_{x \in G} u_x = 0$ and $\sum_{y \in H} v(y) = 0$, then the variance[1] of $\mathbf{t}$ for wrong keys is

$$\frac{|G|^2}{q} \sum_{i=1}^{q} \mathbb{E}|u_{\mathbf{x}_i}|^2 |v(\mathbf{y}_i)|^2 = \left(|G| \sum_{x \in G} |u_x|^2\right)\left(\frac{1}{|H|} \sum_{y \in H} |v(y)|^2\right) = \|u\|_2^2 \|v\|_2^2,$$

where the norms are defined as in Chapter 10.

In words, the product of the lengths of $u$ and $v$ is the standard deviation of the test statistic. One can show, under additional assumptions that are not discussed in detail here, that the standard deviation is approximately the same for the right key. In fact, if we go as far as assuming that the test statistic is normally distributed, then the results in Chapters 4 and 7 imply that the data-complexity is inversely proportional to

$$\frac{|v(T^F u)|}{\|u\|_2 \|v\|_2}.$$

This should, however, be taken with a grain of salt. The assumptions mentioned above do not always hold, so the two-norm is not always the right measure of length. Roughly speaking, although the two-norm in some cases captures the correct local geometry, it does not determine the global geometry.

For the most part, the geometric approach is a combinatorial theory and does not attempt to solve statistical problems such as determining the data-complexity of a given property. However, in some cases choosing a particular norm automatically leads to useful statistical results. Exploring the underlying reasons for this would take us beyond the state of the art.

*Remark* 11.5  As discussed in Chapter 10, the 2-norm is induced by an inner product. This inner product leads to an isometric isomorphism between $\mathbb{C}[G]$ and $\mathbb{C}^G$. In most of this chapter, this isomorphism is not used. However, when working with the 2-norm, thinking of cryptanalytic properties as a pair of subspaces of $\mathbb{C}[G]$ and $\mathbb{C}[H]$ (or $\mathbb{C}^G$ and $\mathbb{C}^H$) sometimes helps with geometric intuition.

---

[1] The variance of a complex-valued random variable $\mathbf{z}$ is $\mathbb{E}\,|\mathbf{z} - \mathbb{E}\mathbf{z}|^2$.

## 11.2 Linear cryptanalysis

The pushforward and pullback operators theoretically make it possible to evaluate cryptanalytic properties (in the sense of Definition 11.1). However, in the standard basis, this is impractical. Most block ciphers involve additions with round keys $k$, for which the pushforward operator is conventionally denoted by $T^k$. Ideally, the properties used in the analysis depend minimally on the key.

For $f$ in $\mathbb{C}^G$, the function $T^{k^\vee} f$ is given by $x \mapsto f(x+k)$. Hence, as discussed in Chapter 10, the Fourier transformation $\mathcal{F}$ diagonalizes all of the matrices $T^{k^\vee}$. Dually, $\mathcal{F}$ diagonalizes the matrices $T^k$. Hence, it makes sense to take the Fourier transformation of all cryptanalytic properties.

### 11.2.1 Correlation matrices

Consider a one-dimensional cryptanalytic property determined by $u$ in $\mathbb{C}[G]$ and $v$ in $\mathbb{C}^H$. The Fourier transformations of $u$ and $v$ are equal to $\widehat{u} = \mathcal{F}_G(u)$ and $\widehat{v} = \mathcal{F}_H(v)$, respectively. The evaluation of the property is given by

$$v(T^\mathsf{F} u) = (\mathcal{F}_H^{-1} \widehat{v})(T^\mathsf{F} \mathcal{F}_G^{-1} \widehat{u}) = \widehat{v}(\mathcal{F}_H T^\mathsf{F} \mathcal{F}_G^{-1} \widehat{u}).$$

The second equality follows from $\mathcal{F}_H = \mathcal{F}_H^{-\vee}$ — see page 144 in Section 10.2.2. The map $\mathcal{F}_H T^\mathsf{F} \mathcal{F}_G^{-1}$ is the Fourier transformation of the pushforward operator $T^\mathsf{F}$.

**Definition 11.6** (Correlation matrix)   Let $\mathsf{F}: G \to H$ be a function between finite commutative groups $G$ and $H$. Define $C^\mathsf{F}: \mathbb{C}[\widehat{G}] \to \mathbb{C}[\widehat{H}]$ as the Fourier transformation of the pushforward operator of $\mathsf{F}$:

$$C^\mathsf{F} = \mathcal{F}_H T^\mathsf{F} \mathcal{F}_G^{-1}.$$

The correlation matrix of $\mathsf{F}$ is the matrix representation of $C^\mathsf{F}$ with respect to the standard bases of $\mathbb{C}[\widehat{G}]$ and $\mathbb{C}[\widehat{H}]$.

The coordinates of the correlation matrix of $\mathsf{F}$ are given by

$$C^\mathsf{F}_{\chi,\psi} = \delta^\chi(C^\mathsf{F}\delta_\psi) = \chi(T^\mathsf{F}\psi^*) = \frac{1}{|G|}\sum_{x \in G}\chi(\mathsf{F}(x))\,\overline{\psi(x)}.$$

These coordinates are evaluations of one-dimensional cryptanalytic properties $(u,v)$ with $u = \psi^*$ and $v = \chi$. Equivalently, $\widehat{u} = \delta_\psi$ and $\widehat{v} = \delta^\chi$.

*Example* 11.3   Linear cryptanalysis uses cryptanalytic properties of the form

$$(U, V) = (\mathrm{Span}\{\psi^*\}, \mathrm{Span}\{\chi\}).$$

The evaluation of this property at unit-norm vectors is $C^\mathsf{F}_{\chi,\psi}$.

If $G = \mathbb{F}_2^n$ and $H = \mathbb{F}_2^m$, then the group characters $\psi$ and $\chi$ are given by $\psi(x) = (-1)^{u^T x}$ and $\chi(x) = (-1)^{v^T x}$. Hence,

$$C^F_{\chi,\psi} = \frac{1}{2^n} \sum_{x \in \mathbb{F}_2^n} (-1)^{u^T x + v^T F(x)}.$$

Up to the isomorphism $u \mapsto (-1)^{u^T x}$ between $\mathbb{F}_2^n$ and $\widehat{\mathbb{F}_2^n}$, and $\mathbb{F}_2^m$ and $\widehat{\mathbb{F}_2^m}$, the coordinates of $C^F$ can be labeled by $u$ and $v$ instead of $\chi$ and $\psi$. This leads back to Definition 2.3.  ▷

The properties of correlation matrices that were proven in Chapter 2 by tedious calculations now become immediate consequences of Theorem 11.4. This is because the properties of the pushforward operator are independent of the choice of basis. In particular, the following result holds.

**Theorem 11.7** *Let* $F: G \to H$ *be a function. The map* $C^F$ *has the following properties:*

(1) *If* $F((x_1, \ldots, x_n)) = (F_1(x_1), \ldots, F_n(x_n))$ *with* $F_i: G_i \to H_i$ *so that* $G = \bigoplus_{i=1}^n G_i$ *and* $H = \bigoplus_{i=1}^n H_i$, *then* $C^F = C^{F_1} \otimes C^{F_2} \otimes \cdots \otimes C^{F_n}$.
(2) *If* $F = F_r \circ \cdots \circ F_2 \circ F_1$, *then* $C^F = C^{F_r} \cdots C^{F_2} C^{F_1}$.

*Proof* Like for Theorem 11.4, it is sufficient to prove the result for $n = 2$ and $r = 2$. The first property relies on the fact that $\mathcal{F}_{G_1 \oplus G_2} = \mathcal{F}_{G_1} \otimes \mathcal{F}_{G_2}$. This is due to Theorem 10.12. Specifically,

$$C^F = (\mathcal{F}_{H_1} \otimes \mathcal{F}_{H_2})(T^{F_1} \otimes T^{F_2})(\mathcal{F}_{G_1} \otimes \mathcal{F}_{G_2})^{-1} = \underbrace{\mathcal{F}_{H_1} T^{F_1} \mathcal{F}_{G_1}^{-1}}_{C^{F_1}} \otimes \underbrace{\mathcal{F}_{H_2} T^{F_2} \mathcal{F}_{G_2}^{-1}}_{C^{F_2}}.$$

Explicitly, the second property follows from Theorem 11.4 (2) as follows:

$$C^{F_2} C^{F_1} = (\mathcal{F}_H T^{F_2} \mathcal{F}_K^{-1})(\mathcal{F}_K T^{F_1} \mathcal{F}_G^{-1}) = \mathcal{F}_H T^F \mathcal{F}_G^{-1} = C^F,$$

where $\mathcal{F}_K$ is the Fourier transformation for the intermediate group $K$.  □

Theorem 2.6 generalizes as follows.

**Theorem 11.8** *Let* $F: G \to H$ *be defined by* $F(x) = L(x) + t$ *with* $L: G \to H$ *a group homomorphism and* $t$ *in* $H$. *The correlation matrix of* $F$ *satisfies*

$$C^F_{\chi,\psi} = \chi(t) \delta^{\chi \circ L}(\psi).$$

*Proof* The matrix $C^t$ is diagonal with the eigenvalues $\chi(t)$ on the diagonal. By the definition of $C^L$, it holds that

$$C^L_{\chi,\psi} = (\chi \circ L)(\psi^*) = \langle \psi, \chi \circ L \rangle = \delta^{\chi \circ L}(\psi).$$

## 11.2 Linear cryptanalysis

The last equality is due to the orthogonality of characters (Theorem 10.13). The overall result follows from $C^F = C^t\, C^L$. □

### 11.2.2 Multiple linear cryptanalysis

Multiple linear cryptanalysis relies on cryptanalytic properties $(U, V)$ with

$$U = \mathsf{Span}\{\psi_1^*, \psi_2^*, \ldots, \psi_n^*\},$$
$$V = \mathsf{Span}\{\chi_1, \chi_2, \ldots, \chi_m\}.$$

In Chapter 6, it was shown that multidimensional linear approximations are special because their correlations characterize the probability distribution of a linear projection of the plaintext-ciphertext pairs for uniform random inputs.

Multidimensional linear cryptanalysis generalizes to arbitrary finite Abelian groups as the special case of multiple linear cryptanalysis where $X^1 = \{\psi_1, \ldots, \psi_n\}$ and $Y^1 = \{\chi_1, \ldots, \chi_m\}$ are subgroups of $\widehat{G}$ and $\widehat{H}$, respectively.

Let $\pi_Y : H \to H/Y$ be the projection defined by $\pi_Y(h) = h + Y$. By Theorem 10.18, the subspace $V = \mathsf{Span}\{\chi \mid \chi \in Y^1\}$ satisfies the following equality:

$$V = \mathsf{Span}\{f \circ \pi_Y \mid f \in \mathbb{C}^{H/Y}\} = \mathsf{Span}\{\delta^{h+Y} \mid h + Y \in H/Y\},$$

where $\delta^{h+Y} = \sum_{y \in h+Y} \delta^y$ like in Example 11.1. As discussed at the end of Section 10.2.3, there is a similar equality for the subspace $U = \mathsf{Span}\{\psi^* \mid \psi \in X^1\}$. It is obtained by applying the anti-isomorphism $x \mapsto x^*$ from Theorem 10.6:

$$U = \mathsf{Span}\{\delta_{g+X} \mid g + X \in G/X\}.$$

The result of evaluating $(U, V)$ at $\delta_{g+X}/|X|$ in $U$ and $\delta^{h+Y}$ in $V$ is

$$\frac{1}{|X|} \sum_{x \in g+X} \delta^{h+Y}(\mathsf{F}(x) + Y) = \Pr_{\mathbf{x}}\left[\mathsf{F}(\mathbf{x}) \equiv h \bmod Y\right],$$

with $\mathbf{x}$ uniform random on the coset $g + X$. This is the same probability as in Corollary 6.6. From the equalities of the subspaces above, it follows that this probability can be expressed in terms of the correlations of linear approximations with characters in $X^1$ and $Y^1$. Exercise 11.4 asks you to make this explicit.

## 11.3 Exact propagation

In the first part of Chapter 9, "exact" extensions of linear cryptanalysis were discussed. Examples include saturation properties, zero-correlation linear approximations and invariants.

### 11.3.1 Forward

To propagate a subspace $U$ of $\mathbb{C}[G]$ through a function $\mathsf{F}\colon G \to H$ means to determine the image $T^\mathsf{F} U$. This is generally not feasible, which is why cryptanalytic properties are used instead, but in some cases it is possible to show that $T^\mathsf{F} U \subseteq W$ for a subspace $W$ of $\mathbb{C}[H]$.

*Example* 11.4  If $\mathsf{F}(X) \subseteq Y$ for subsets $X$ and $Y$ of $G$ and $H$, respectively, then
$$T^\mathsf{F} \operatorname{Span}\{\delta_x \mid x \in X\} \subseteq \operatorname{Span}\{\delta_y \mid y \in Y\}.$$
This inclusion implies that the cryptanalytic property $(U, V)$ with $U = \operatorname{Span}\{\delta_X\}$ and $V = \operatorname{Span}\{\delta^Y\}$ evaluates to $\delta^Y(T^\mathsf{F} \delta_X) = |X|$. ▷

Not every exact propagation corresponds to an inclusion of sets.

*Example* 11.5  The method from Section 9.1.1 to find saturation properties is based on describing the state as one of several possible sets, without saying what happens to the individual elements of those sets. That is, for families of sets $S_1, \ldots, S_n$ and $T_1, \ldots, T_m$ with $m \geq n$,
$$T^\mathsf{F} \operatorname{Span}\{\delta_{S_1}, \ldots, \delta_{S_n}\} \subseteq \operatorname{Span}\{\delta_{T_1}, \ldots, \delta_{T_m}\}.$$
For example, for $\mathsf{P}\colon H \to Y$ a "projection function" (see Section 9.2.3), the sets $T_1, \ldots, T_m$ could be characterized by $T^\mathsf{P} \delta_{T_i} \in \operatorname{Span}\{\delta_Y\}$. Let $(U, V)$ be a property with $U = \operatorname{Span}\{\delta_{S_i}\}$ and $V = \operatorname{Span}\{\delta^y \circ \mathsf{P} \mid y \in Y\}$. By the above inclusion, the evaluations of the property $(U, V)$ satisfy
$$(\delta^y \circ \mathsf{P})(T^\mathsf{F} \delta_{S_i}) = \delta^y(T^\mathsf{P} T^\mathsf{F} \delta_{S_i}) = \frac{|S_i|}{|Y|}.$$
That is, $\mathsf{P} \circ \mathsf{F}$ is saturated on the input set $S_i$. ▷

### 11.3.2 Backward

In addition to propagating a subspace of $\mathbb{C}[G]$ "forward," as in Section 11.3.1, it is possible to propagate a subspace $V$ of $\mathbb{C}^H$ backward through a function $\mathsf{F}\colon G \to H$. This amounts to showing $T^{\mathsf{F}^\vee} V \subseteq W$ for a subspace $W$ of $\mathbb{C}^G$.

*Example* 11.6  Let $f\colon G \to X$ and $g\colon H \to Y$ be two functions. This generalizes the concept of nonlinear approximations from Section 9.2.2. If

$$T^{\mathsf{F}^\vee} \mathsf{Span}\{\delta^y \circ g \mid y \in Y\} \subseteq \mathsf{Span}\{\delta^x \circ f \mid x \in X\},$$

then there exists a function $h\colon X \to Y$ such that $h \circ f = g \circ \mathsf{F}$. ▷

### 11.3.3 Zero-correlation

A property $(U, V)$ that evaluates to zero for all $u$ in $U$ and $v$ in $V$ is called a zero-correlation property. Let $\mathsf{F} = \mathsf{F}_2 \circ \mathsf{F}_1$ with $\mathsf{F}_1\colon G \to H$ and $\mathsf{F}_2\colon H \to K$. The annihilator of a subspace $W$ of $\mathbb{C}^H$ is defined as (see also Exercise 10.6)

$$W^0 = \{x \in \mathbb{C}[H] \mid \forall w \in W\colon w(x) = 0\}.$$

If $T^{\mathsf{F}_1} U \subseteq W^0$ and $T^{\mathsf{F}_2^\vee} V \subseteq W$ for some subspace $W$, then $(U, V)$ is a zero-correlation property:

$$v(T^{\mathsf{F}} u) = \left(T^{\mathsf{F}_2^\vee} v\right)\left(T^{\mathsf{F}_1} u\right) = 0$$

for all $u$ in $U$ and $v$ in $V$. This generalizes the miss-in-the-middle principle from Section 8.2 that is used to find zero-correlation linear approximations.

*Example* 11.7  Let $S_1, \ldots, S_n$ and $T_1, \ldots, T_m$ be two families of sets like in Example 11.5, so that $T^{\mathsf{P}} \delta_{T_i} \in \mathsf{Span}\{\delta_Y\}$ for a function $\mathsf{P}\colon H \to Y$. As before, assume that $T^{\mathsf{F}} \mathsf{Span}\{\delta_{S_1}, \ldots, \delta_{S_n}\} \subseteq \mathsf{Span}\{\delta_{T_1}, \ldots, \delta_{T_m}\}$. Let $U = \mathsf{Span}\{\delta_{S_i}\}$ and, for an arbitrary constant $z$ in $Y$,

$$V = \mathsf{Span}\{\delta^y \circ \mathsf{P} - \delta^z \circ \mathsf{P} \mid y \in Y\}.$$

Since $V^0 \supseteq \mathsf{Span}\{\delta_{T_1}, \ldots, \delta_{T_m}\}$, the property $(U, V)$ is zero-correlation. ▷

### 11.3.4 Invariants

An invariant of a function $\mathsf{F}\colon G \to G$ is a subspace $U$ of $\mathbb{C}[G]$ so that $T^{\mathsf{F}} U \subseteq U$.

*Example* 11.8  Recall from Chapter 9 that a nonlinear invariant of a function $\mathsf{F}\colon \mathbb{F}_2^n \to \mathbb{F}_2^n$ is a function $f\colon \mathbb{F}_2^n \to \mathbb{F}_2$ such that there exists a constant $c$ so that $f(\mathsf{F}(x)) = f(x) + c$ for all $x$ in $\mathbb{F}_2^n$. Equivalently, if $S = \{x \in \mathbb{F}_2^n \mid f(x) = 1\}$, then either $\mathsf{F}(S) \subseteq S$ and $\mathsf{F}(\mathbb{F}_2^n \setminus S) \subseteq \mathbb{F}_2^n \setminus S$ or $\mathsf{F}(S) \subseteq \mathbb{F}_2^n \setminus S$ and $\mathsf{F}(\mathbb{F}_2^n \setminus S) \subseteq S$. The following vector space $U$ corresponds to this invariant:

$$U = \mathsf{Span}\{\delta_S, \delta_{\mathbb{F}_2^n \setminus S}\}.$$

Alternatively, the following subspace $V$ of $\mathbb{C}^G$ satisfies $T^{F^\vee} V \subseteq V$:

$$V = \mathsf{Span}\{\delta^0 \circ f, \delta^1 \circ f\}.$$

The space $V$ consists of all complex-valued functions on $\mathbb{F}_2$, pulled back to $\mathbb{F}_2^n$ along $f$. To test if the invariant holds, one can evaluate the property $(U, V)$. ▷

To distinguish between the spaces $U$ and $V$ in Example 11.8, $U$ may be called a forward invariant and $V$ a backward invariant. Verify that if $U$ is a forward invariant, then $U^0$ is a backward invariant and conversely. Furthermore, using the anti-isomorphism $x \mapsto x^*$ from Theorem 10.6, $U$ is a forward invariant if and only if $U^*$ is a backward invariant.

Invariants are related to the eigenvectors of $T^F$.

**Theorem 11.9** *Every invariant $U$ of a permutation $\mathsf{F}: G \to G$ has a basis consisting of eigenvectors of $T^F$.*

*Proof* The pushforward operator $T^F$ is diagonalizable. Indeed, since $\mathsf{F}^n$ is the identity function for some $n \geq 1$, the minimal polynomial of $T^F$ divides $x^n - 1$. This polynomial has distinct roots over $\mathbb{C}$.

If $U$ is an invariant and $\mathsf{F}$ a permutation, then $T^F U = U$. It follows that the minimal polynomial of the restriction $T^F|_U : U \to U$ divides the minimal polynomial of $T^F$. Hence, $T^F|_U$ is diagonalizable. □

Theorem 11.9 implies that the Fourier transformation of an invariant has a basis consisting of eigenvectors of $C^F$.

*Example* 11.9 The invariant $U$ from Example 11.8 is the span of two eigenvectors of $T^F$:

$$U = \mathsf{Span}\{\delta_S + \delta_{\mathbb{F}_2^n \setminus S}, \delta_S - \delta_{\mathbb{F}_2^n \setminus S}\}.$$

The first vector is equal to $\delta_S + \delta_{\mathbb{F}_2^n \setminus S} = \delta_{\mathbb{F}_2^n}$. In fact, this is an eigenvector of $T^F$ for all permutations $\mathsf{F}$. Similarly, $V$ is the span of two eigenvectors of $T^{F^\vee}$:

$$V = \mathsf{Span}\{x \mapsto 1, x \mapsto (-1)^{f(x)}\}.$$

In particular, the Fourier transformation of $(-1)^f$ is an eigenvector of $C^{F^\vee}$. ▷

## 11.4 Approximate propagation

The evaluations of every cryptanalytic property determine a linear function, called its approximation map. This map is used to glue together a sequence of properties of functions to a property of their composition.

## 11.4 Approximate propagation

### 11.4.1 Approximation maps

The data of all evaluations $v(T^F u)$ of a cryptanalytic property $(U, V)$ are equivalent to the linear map $U \to \mathbb{C}[H]/V^0$ defined by

$$u \mapsto T^F u + V^0.$$

Here, $V^0$ is the annihilator of the subspace $V$. That is, the state is only known up to addition by a vector that cannot be "measured" using functions in $V$.

The above map might as well be called the approximation map of $(U, V)$, but the problem is that to be able to compose the map $U \to \mathbb{C}[H]/V^0$ with the map of another property, an embedding of $\mathbb{C}[H]/V^0$ into $\mathbb{C}[H]$ is needed. Fortunately, choosing an algebraic complement of $V$ solves this problem. The remainder of this chapter uses orthogonal complements:[2] $\mathbb{C}^H = V \oplus V^\perp$. By Exercise 10.6 and with $x \mapsto x^*$ denoting the anti-isomorphism from Theorem 10.6, $(V^\perp)^0 = V^*$. Hence, since $V^0 \cap (V^\perp)^0 = (V \oplus V^\perp)^0 = \{0\}$,

$$\mathbb{C}[H] = V^0 \oplus V^*.$$

The point of all this is that $\mathbb{C}[H]/V^0$ can be identified with $V^*$ to obtain a map $U \to V^*$. Importantly, $V^*$ is a subspace of $\mathbb{C}[H]$. Beware that in the following definition and the discussion below, the property is $(U, V^*)$ rather than $(U, V)$ to simplify notation. This is reasonable because, for all intents and purposes, $V^{**} = V$.

**Definition 11.10** The approximation map of a cryptanalytic property $(U, V^*)$ of a function $\mathsf{F}\colon G \to H$ is the linear map $\langle V, U \rangle_\mathsf{F}\colon U \to V$ defined by

$$\langle V, U \rangle_\mathsf{F} = \pi_V \, T^\mathsf{F} \, \iota_U,$$

with $\iota_U\colon U \to \mathbb{C}[G]$ inclusion and $\pi_V\colon \mathbb{C}[H] \to V$ orthogonal projection.

The idea behind the approximation map is that it sends $u$ in $U$ to a vector in $V$ that approximates $T^\mathsf{F} u$ in the following sense. For all $u$ in $U$ and $v$ in $V^*$,

$$v\bigl(\langle V, U \rangle_\mathsf{F} u\bigr) = v\bigl(\pi_V T^\mathsf{F} \iota_U u\bigr) = v\bigl(T^\mathsf{F} u\bigr).$$

The last equality follows from $\iota_U u = u$ and $T^\mathsf{F} u = \pi_V T^\mathsf{F} u + \pi_{V^\perp} T^\mathsf{F} u$, along with the fact that $V^*$ is the annihilator of $V^\perp$. In other words, replacing $T^\mathsf{F}$ by its approximation map does not affect evaluations of the property.

### 11.4.2 Geometry

Due to the best approximation theorem (Theorem 10.8), the approximations obtained by applying $\langle V, U \rangle_\mathsf{F}$ are geometrically best possible. The quality of a cryptanalytic property is measured by its *principal correlations*.

---

[2] This "arbitrary" choice can be avoided using a slight generalization of Definition 11.1.

**Definition 11.11** (Principal correlations)  Let $(U, V^*)$ be a cryptanalytic property of a function $\mathsf{F}\colon G \to H$. Let $d = \min\{\dim U, \dim V\}$. The principal correlations of the property $(U, V^*)$ are the $d$ largest singular values of its approximation map $\langle V, U \rangle_\mathsf{F}$.

If $\mathsf{F}$ is injective, then the principal correlations of $(U, V^*)$ are equal to the cosines of the $d$ smallest principal angles between the subspaces $T^\mathsf{F} U$ and $V$ (see Exercise 10.7). The principal correlation of a linear approximation $(\operatorname{Span}\{\psi^*\}, \operatorname{Span}\{\chi\})$ is the absolute value of its correlation.

The principal correlations also have a statistical interpretation. Without going into details, if $\sigma_1, \ldots, \sigma_r$ are the first $r$ principal correlations of a property, then (under assumptions that we leave out here) the minimal data-complexity of a hypothesis test based on known-plaintext estimates of at most $r$ evaluations of the property is inversely proportional to

$$\sum_{i=1}^{r} \sigma_i^2.$$

The sum of the squares of all principal correlations of a multiple linear approximation is equal to its capacity.

### 11.4.3 Principle of dominant trails

Suppose that $\mathsf{F} = \mathsf{F}_r \circ \cdots \circ \mathsf{F}_1$ with $\mathsf{F}_i \colon G_i \to G_{i+1}$. For linear cryptanalysis, the multiplication property of correlation matrices leads to the concept of linear trails $(\chi_1, \ldots, \chi_{r+1})$:

$$c^\mathsf{F}_{\chi_{r+1}, \chi_1} = \sum_{\chi_2, \ldots, \chi_r} \prod_{i=1}^{r} c^{\mathsf{F}_i}_{\chi_{i+1}, \chi_i}.$$

In the above, the correlations $c^\mathsf{F}_{\chi_{i+1}, \chi_i}$ are evaluations of cryptanalytic properties of the functions $\mathsf{F}_1, \ldots, \mathsf{F}_r$.

The following result shows that there is a similar expression for the approximation map of arbitrary properties of $\mathsf{F}$. A sequence of vector spaces $(U_1, U_2, \ldots, U_{r+1})$ or, equivalently, of compatible cryptanalytic properties $(U_1, U_2^*), (U_2, U_3^*), \ldots, (U_r, U_{r+1}^*)$, is called a trail.

**Theorem 11.12**  For $1 \le i \le r+1$, let $\Omega_i$ be a set of orthogonal subspaces of $\mathbb{C}[G_i]$ such that $\mathbb{C}[G_i] = \bigoplus_{U \in \Omega_i} U$. For every property $(U_1, U_{r+1}^*)$ of $\mathsf{F}$ with $U_1$ in $\Omega_1$ and $U_{r+1}$ in $\Omega_{r+1}$,

$$\langle U_{r+1}, U_1\rangle_{\mathsf{F}} = \sum_{U_2,\ldots,U_r} \langle U_{r+1}, U_r\rangle_{\mathsf{F}_r} \cdots \langle U_3, U_2\rangle_{\mathsf{F}_2} \langle U_2, U_1\rangle_{\mathsf{F}_1},$$

where the sum is over all $(U_2, \ldots, U_r)$ in $\prod_{i=2}^{r} \Omega_i$.

*Proof* By definition, $\langle U_{r+1}, U_i\rangle_{\mathsf{F}_r \circ \cdots \circ \mathsf{F}_i} = \pi_{U_{r+1}} T^{\mathsf{F}_r \circ \cdots \circ \mathsf{F}_i} \iota_{U_i}$. Furthermore, by the definition of $\Omega_{i+1}$, the map $\sum_{U \in \Omega_{i+1}} \pi_U$ is the identity. Hence,

$$\langle U_{r+1}, U_i\rangle_{\mathsf{F}_r \circ \cdots \circ \mathsf{F}_i} = \sum_{U_{i+1} \in \Omega_{i+1}} \langle U_{r+1}, U_{i+1}\rangle_{\mathsf{F}_r \circ \cdots \circ \mathsf{F}_{i+1}} \langle U_{i+1}, U_i\rangle_{\mathsf{F}_i}.$$

The result follows by repeatedly applying this equality for $i = 1, \ldots, r-1$. □

Since knowing the approximation map of a property is equivalent to knowing all of its evaluations, Theorem 11.12 provides a way to glue together properties of $\mathsf{F}_1, \ldots, \mathsf{F}_r$. In practice, the sum over all trails is approximated by summing over a small set of dominant trails.

## 11.5 Historical remarks

The starting point of the geometric approach was the idea that correlation matrices represent linear maps. If one takes this point of view seriously, then it is important to understand the vector spaces on which they act. The first application of this point of view was the analysis of invariants in Beyne (2018). Linear cryptanalysis and its extensions were discussed in Beyne (2021).

The case of linear cryptanalysis serves as an introduction to the geometric approach in general, which is important to understand other cryptanalytic techniques, such as differential and integral cryptanalysis and the relations between them Beyne (2023).

## 11.6 References

Beyne, Tim (Dec. 2018). "Block Cipher Invariants as Eigenvectors of Correlation Matrices." In: *ASIACRYPT 2018, Part I*. Ed. by Thomas Peyrin and Steven Galbraith. Vol. 11272. LNCS. Springer, Cham, pp. 3–31. DOI: 10.1007/978-3-030-03326-2_1.
— (Dec. 2021). "A Geometric Approach to Linear Cryptanalysis." In: *ASIACRYPT 2021, Part I*. Ed. by Mehdi Tibouchi and HuaxiongWang. Vol. 13090. LNCS. Springer, Cham, pp. 36–66. DOI: 10.1007/978-3-030-92062-3_2.
— (June 2023). "A Geometric Approach to Symmetric-key Cryptanalysis." PhD thesis. KU Leuven.

## 11.7 Exercises

### Exercise 11.1

Let $G$ be a finite Abelian group. A fixed point of a function $F\colon G \to G$ is an element $x$ of $G$ such that $F(x) = x$. Recall that the trace $\operatorname{Tr} A$ of a matrix $A$ is the sum of its diagonal entries. Prove that $\operatorname{Tr} C^F$ is equal to the number of fixed points of $F$.

### Exercise 11.2

Let $F\colon G \to H$ be a function. A pair of inputs $(x, y)$ is called a collision for $F$ if $F(x) = F(y)$. Prove the following formula for the probability that a uniform random pair of inputs is a collision:

$$|H| \Pr_{\mathbf{x},\mathbf{y}}[F(\mathbf{x}) = F(\mathbf{y})] = \sum_{\chi \in \widehat{H}} |C^F_{\chi,1}|^2.$$

In the above, $\mathbf{x}$ and $\mathbf{y}$ are uniform random variables on $G$ and $C^F$ is the correlation matrix of $F$.

### Exercise 11.3

Let $\mathbb{F}_q$ be a finite field of order $q$ and $f$ a polynomial over $\mathbb{F}_q$ of degree $d \geq 2$ coprime to $q$. One of the consequences of the Riemann hypothesis for curves over finite fields is the following exponential sum estimate:

$$\left| \sum_{x \in \mathbb{F}_q} \exp\left(\frac{2\pi i \operatorname{Tr} f(x)}{p}\right) \right| \leq (d-1)\sqrt{q}.$$

This result is known as Weil's bound.

1. Let $F\colon \mathbb{F}_q \mapsto \mathbb{F}_q$ be the cube function defined by $F(x) = x^3$. Prove that

$$|C^F_{\chi,\psi}| \leq 2/\sqrt{q}$$

for all nontrivial characters $\chi$ and assuming that $q$ is not a power of three.

2. Suppose that $q \equiv 2 \pmod{3}$. Let $G\colon \mathbb{F}_q \mapsto \mathbb{F}_q$ be the function defined by $G(x) = (x^3 + k)^{1/3}$. Prove that if $k \neq 0$ and $q$ is odd, then

$$|C^G_{\chi,\psi}| \leq 2/\sqrt{q}$$

for all nontrivial characters $\chi$.

3. What goes wrong in the second question when $q$ is even?

### Exercise 11.4

Let $G$ and $H$ be finite Abelian groups with subgroups $X$ and $Y$, respectively, like in Section 11.2.2. Let $F\colon G \to H$ be a function and consider the probabilities

$$\Pr_{\mathbf{x}}\left[F(\mathbf{x}) \equiv h \bmod Y\right],$$

with $\mathbf{x}$ uniform random on the coset $g + X$. By Theorem 10.18, these probabilities can be expressed as linear combinations of the correlations $C^{\mathsf{F}}_{\chi,\psi}$ with $\psi$ in $X^1$ and $\chi$ in $Y^1$.

1. Prove the following equality for $\mathbf{x}$ uniform random on $g + X$:

$$\Pr_{\mathbf{x}}\left[F(\mathbf{x}) \equiv h \bmod Y\right] = \frac{|Y|}{|H|} \sum_{\substack{\psi \in X^1 \\ \chi \in Y^1}} \overline{\chi(h)}\, \psi(g)\, C^{\mathsf{F}}_{\chi,\psi}.$$

   Avoid lengthy calculations like those in the proof of Theorem 6.3.

2. Prove the inverse relationship: Write $C^{\mathsf{F}}_{\chi,\psi}$ (with $\psi$ in $X^1$ and $\chi$ in $Y^1$) as a linear combination of the probabilities for all cosets $g + X$ and $h + Y$.

### Exercise 11.5

Give a function $F\colon G \to G$ such that $T^{\mathsf{F}}$ is not diagonalizable.

### * Exercise 11.6

Let $G$ be a finite Abelian group and $F\colon G \to G$ a permutation. Suppose that $F$ has $\ell$ disjoint cycles of lengths $l_1, \ldots, l_\ell$, with cycle $i$ consisting of values $(x_{i,1}, x_{i,2}, \ldots, x_{i,l_i})$.

1. What are the eigenvalues of $T^{\mathsf{F}}$?
2. Give the corresponding eigenvectors of $T^{\mathsf{F}}$ and $C^{\mathsf{F}}$.
3. Describe a permutation $F\colon \mathbb{F}_2^{1337} \to \mathbb{F}_2^{1337}$ without nontrivial invariant subspaces.

### Exercise 11.7

Let $\mathsf{P}_1\colon G \to X$ and $\mathsf{P}_2\colon H \to Y$ be balanced "projections," and let $F\colon G \to H$ be a function. Note that $\mathsf{P}_1$ induces a partition of $G$ as follows:

$$G = \bigcup_{x \in X} \mathsf{P}_1^{-1}(x).$$

In the following, let $u = \delta_{\mathsf{P}_1^{-1}(x)}$ and $v = \delta^y \circ \mathsf{P}_2$ for $x$ in $X$ and $y$ in $Y$.

1. Prove the following equality:
$$v(T^F u) = |\{z \in G \mid P_1(z) = x \wedge P_2(F(z)) = y\}|$$
for all $x$ in $X$ and all $y$ in $Y$.

2. Deduce that the coordinates of the approximation map of (Span$\{u\}$, Span$\{v\}$) relative to normalized bases are given by
$$\Pr_z[P_2(F(\mathbf{z})) = y \mid P_1(\mathbf{z}) = x],$$
where the probability is over a uniform random $\mathbf{z}$.

3. Suppose that $G = \mathbb{F}_2^n$, $X = Y = \mathbb{F}_2$, $P_1(x) = u^\mathsf{T} x$ and $P_2(x) = v^\mathsf{T} x$. Show that if F is a permutation, then there exist bases so that the matrix representation of $\langle V, U \rangle_\mathsf{F}$ is
$$\begin{bmatrix} 1 & 0 \\ 0 & c \end{bmatrix},$$
where $c$ is the correlation of the linear approximation $(u, v)$. What are the principal correlations of $(U, V)$?

## Exercise 11.8

Let $(U, V)$ be a multiple linear property of $\mathsf{F}: G \to G$ that contains the trivial linear approximation. The square of the Frobenius norm $\| \cdot \|_{\mathrm{fr}}^2$ of a matrix is the sum of the squares of its singular values.

1. Prove that $\|\langle V, U \rangle_\mathsf{F}\|_{\mathrm{fr}}^2 - 1$ is the capacity of the multiple linear approximation.

2. Use the fact that the Frobenius norm does not change under unitary change of basis to show that $\|\langle V, U \rangle_\mathsf{F}\|_{\mathrm{fr}}^2 - 1$ is equal to the squared Euclidean imbalance.

## Exercise 11.9

A *coalgebra* over $\mathbb{C}$ is a vector space $V$ with a *coproduct* $\Delta: V \to V \otimes V$ that satisfies a number of axioms. For example, $\mathbb{C}[G]$ is a coalgebra with coproduct
$$\Delta(\delta_x) = \delta_{(x,x)}. \tag{11.1}$$

The following questions are about the map $\Delta: \mathbb{C}[G] \to \mathbb{C}[G^2]$ defined by (11.1). Let id denote the identity function on $\mathbb{C}[G]$.

1. Prove that $\Delta$ is *coassociative*: $(\mathrm{id} \otimes \Delta) \circ \Delta = (\Delta \otimes \mathrm{id}) \circ \Delta$.

2. Show that there exists a *counit* $\varepsilon: \mathbb{C}[G] \to \mathbb{C}$ satisfying $(\mathrm{id} \otimes \varepsilon) \circ \Delta = \mathrm{id}$ and $(\varepsilon \otimes \mathrm{id}) \circ \Delta = \mathrm{id}$.

3. A morphism of coalgebras is a linear map $T: \mathbb{C}[G] \to \mathbb{C}[H]$ that satisfies $\Delta_H \circ T = (T \otimes T) \circ \Delta_G$, with $\Delta_G$ the coproduct on $\mathbb{C}[G]$ and $\Delta_H$ the coproduct on $\mathbb{C}[H]$. Prove that for every morphism of coalgebras $T: \mathbb{C}[G] \to \mathbb{C}[H]$, there exists a function $\mathsf{F}: G \to H$ such that $T = T^{\mathsf{F}}$.

The coalgebra structure of $\mathbb{C}[G]$ plays an important role in the geometric approach to cryptanalysis in general.

# Appendix A  Normal distribution

This appendix collects some import facts about the normal distribution. These results are used throughout this book, and in particular in Chapters 4, 6 and 7.

## A.1 Univariate normal distribution

Normal distributions are a family of continuous probability distributions. The *standard normal distribution* is the distribution with density function

$$\varphi(x) = \frac{1}{\sqrt{2\pi}} e^{-\frac{1}{2}x^2}.$$

This probability density function is illustrated in Figure A.1.

The cumulative distribution function of the standard normal distribution is denoted by $\Phi$. By definition, it is equal to

$$\Phi(x) = \int_{-\infty}^{x} \varphi(z)\, dz.$$

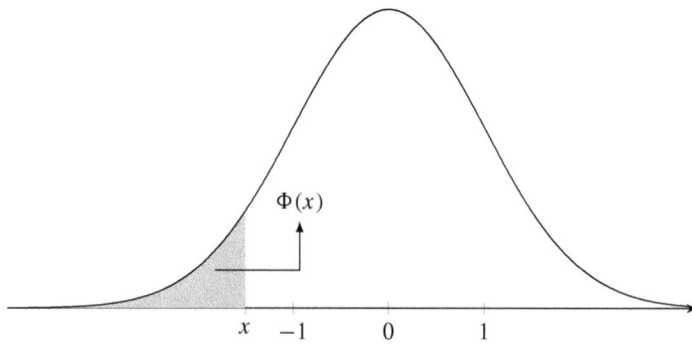

Figure A.1  Probability density function of the standard normal distribution.

It follows from the symmetry of $\varphi$ that the mean of the standard normal distribution is zero. Using integration by parts, it can be shown that the variance is equal to one.

The other normal distributions are obtained from the standard normal distribution by scaling and translation. If $\mathbf{x}$ is random variable following the standard normal distribution, then the cumulative distribution function of $\sigma \mathbf{x} + \mu$ is the function that maps $x$ to

$$\Phi\left(\frac{x-\mu}{\sigma}\right).$$

Equivalently, the density function of $\sigma \mathbf{x} + \mu$ maps $x$ to

$$\frac{1}{\sqrt{2\pi}\sigma} e^{-\frac{1}{2}\left(\frac{x-\mu}{\sigma}\right)^2}.$$

The mean and variance uniquely determine each member of the normal family. Hence, it is reasonable to denote the normal distribution with mean $\mu$ and variance $\sigma^2$ by $\mathcal{N}(\mu, \sigma^2)$.

The normal distribution is of singular importance due its connection with the following limit theorem.

**Theorem A.1** (Central limit theorem)  *Let $\mathbf{x}_1, \mathbf{x}_2, \ldots$ be a sequence of independent random variables on $\mathbb{R}$ with mean $\mu$ and variance $\sigma^2$. In the limit as $n \to \infty$, the distribution of $\sum_{i=1}^{n} \mathbf{x}_i / \sqrt{n}$ converges to $\mathcal{N}(\mu, \sigma^2)$.*

One of the consequences of Theorem A.1 is the normal approximation to the binomial distribution. Specifically, the binomial distribution with $n$ trials of probability $p$ is well approximated by $\mathcal{N}(np, np(1-p))$ for large $n$.

## A.2 Multivariate normal distribution

The standard multivariate normal distribution is the probability distribution of a vector of $d$ independent standard normal distributions. Consequently, its density is given by

$$\varphi(x) = \prod_{i=1}^{d} \frac{1}{\sqrt{2\pi}} e^{-\frac{1}{2}x_i^2} = \frac{1}{\sqrt{(2\pi)^d}} e^{-\frac{1}{2}\|x\|_2^2},$$

with $(x_1, \ldots, x_d)$ the coordinates of the vector $x$ in $\mathbb{R}^d$.

Other multivariate normal distributions are obtained from the standard one by affine transformations. Specifically, if $\mathbf{x}$ is a random variable following the standard multivariate normal distribution, then the probability density function of $A\mathbf{x} + \mu$ is the function that maps $x$ to

$$\frac{1}{|\det A|} \frac{1}{\sqrt{(2\pi)^d}} e^{-\frac{1}{2}\|A^{-1}(x-\mu)\|_2^2},$$

with $\mu$ a vector in $\mathbb{R}^d$ and $A$ an invertible $d \times d$ matrix. The factor $|\det A|$ is the Jacobian determinant of the transformation $x \mapsto Ax + \mu$. The vector $\mu$ is the mean of $A\mathbf{x} + \mu$. The covariance matrix of $A\mathbf{x} + \mu$ is equal to

$$\Sigma = \mathbb{E}_{\mathbf{x}}(A\mathbf{x})(A\mathbf{x})^\mathsf{T} = A\left(\mathbb{E}_{\mathbf{x}}\mathbf{x}^\mathsf{T}\mathbf{x}\right)A^\mathsf{T} = AA^\mathsf{T}.$$

The mean $\mu$ and the covariance matrix $\Sigma$ uniquely identify every member of the multivariate normal family. The multivariate normal distribution with mean $\mu$ and covariance matrix $\Sigma$ is denoted by $\mathcal{N}(\mu, \Sigma)$. Its probability density function maps $x$ in $\mathbb{R}^d$ to

$$\frac{1}{\sqrt{|\det \Sigma|}} \frac{1}{\sqrt{(2\pi)^d}} e^{-\frac{1}{2}(x-\mu)^\mathsf{T}\Sigma^{-1}(x-\mu)}.$$

If $\mathbf{x}$ is a random variable with distribution $\mathcal{N}(\mu, \Sigma)$, then for all vectors $v$ in $\mathbb{R}^d$, the distribution of $v^\mathsf{T}\mathbf{x}$ is the univariate normal distribution $\mathcal{N}(v^\mathsf{T}\mu, v^\mathsf{T}\Sigma v)$. The mean follows by linearity of expectation, and the variance from

$$\mathbb{V}_{\mathbf{x}} v^\mathsf{T}\mathbf{x} = \mathbb{E}_{\mathbf{x}}\left(v^\mathsf{T}(\mathbf{x}-\mu)\right)^2 = v^\mathsf{T}\left(\mathbb{E}_{\mathbf{x}}(\mathbf{x}-\mu)(\mathbf{x}-\mu)^\mathsf{T}\right)v = v^\mathsf{T}\Sigma v.$$

There is a variant of the central limit theorem for multivariate distributions.

**Theorem A.2** (Central limit theorem) *Let $\mathbf{x}_1, \mathbf{x}_2, \ldots$ be a sequence of independent random variables on $\mathbb{R}^d$ with mean $\mu$ and covariance matrix $\Sigma$. In the limit as $n \to \infty$, the distribution of $\sum_{i=1}^n \mathbf{x}_i / \sqrt{n}$ converges to $\mathcal{N}(\mu, \Sigma)$.*

# Appendix B    Statistical formulary

The formulas in Table B.1 rely on several approximations. These are listed in Table B.2. The following specific comments apply to Table B.1:

**Single approximations.** For sampling without replacement, divide the argument of $\Phi$ by $\sqrt{1 - q/2^n}$ (see Section 4.3, page 61).

**Multiple approximations.** If the correlations are known up to sign, replace $\mathrm{Cap}(\Lambda)$ by (see Section 6.1.2, page 80)

$$\sqrt{\sum_{i=1}^{|\Lambda|} c_i^4}\,.$$

**Multidimensional approximations.** If the correlations are unknown and $\Lambda = \Lambda_{\mathsf{in}} \oplus \Lambda_{\mathsf{out}}$, replace $|\Lambda|$ by $|\Lambda_{\mathsf{out}}|$ when chosen plaintexts are available (see Section 6.2.3, page 85). For multidimensional zero-correlation linear approximations, replace $|\Lambda|$ by $|\Lambda_{\mathsf{in}}|^2 |\Lambda_{\mathsf{out}}|$ (see Section 8.4, page 116).

If the correlations or the capacity are key-dependent, use the following formula:

$$\sum_{k \in \mathcal{K}} f_k \, P_{\mathsf{S}}(k),$$

with $f_k$ the frequency of key class $k$. These frequencies are derived from the prior distribution of the key. Additional analysis of the key schedule may be required. The above formula can be adapted to take into account model errors, as discussed in Section 7.3.2 on page 102.

Table B.1. *Basic statistical formulas for the success probability $P_S$*

|  | Correlations | |
|---|---|---|
|  | Known | Unknown |
| Single approximation | $\Phi\left(\lvert c\rvert\sqrt{q} + \Phi^{-1}(P_F)\right)$ | $\Phi\left(\lvert c\rvert\sqrt{q} + \Phi^{-1}(P_F/2)\right)$ |
| Multiple approximations | $\Phi\left(\sqrt{\mathsf{Cap}(\Lambda)\,q} + \Phi^{-1}(P_F)\right)$ | $\Phi\left(\dfrac{\mathsf{Cap}(\Lambda)}{\sqrt{2\lvert\Lambda\rvert}}\,q + \Phi^{-1}(P_F)\right)$ |
| Multiple zero-correlation | $\Phi\left(\dfrac{\sqrt{\lvert\Lambda\rvert}}{2^{n+\frac{1}{2}}}\,q + \Phi^{-1}(P_F)\right)$ | / |

Table B.2. *Approximations used in the formulas from Table B.1*

|  | Correlations | |
|---|---|---|
|  | Known | Unknown |
| Single approximation | correlation zero for wrong keys<br>$q$ large (normal approximation)<br>$c^2 \ll 1$ (constant variance) | |
|  | $c$ fixed | $c^2$ fixed<br>$P_F$ small |
| Multiple approximations | all correlations zero for wrong keys<br>$q/\sqrt{\lvert\Lambda\rvert}$ large (normal approximation)<br>$\lvert C^F_{v_i+v_j,\,u_i+u_j}\rvert \ll 1/\sqrt{\lvert\Lambda\rvert}$<br>whenever $(u_i+u_j, v_i+v_j) \notin \Lambda$<br>$c_i c_j \ll 1$ (negligible covariances) | |
|  | $c_1, c_2 \ldots$ fixed | $\mathsf{Cap}(\Lambda)$ fixed<br>$q c_i^2$ small |

# Appendix C  List of block ciphers

Table C.1. *List of block ciphers mentioned in this book*

| Block cipher | Chapter | Reference |
| --- | --- | --- |
| 3-Way | Chapter 1, page 3 | Joan Daemen, René Govaerts, and Joos Vandewalle (Dec. 1994). "A New Approach to Block Cipher Design." In: *FSE'93*. Ed. by Ross J. Anderson. Vol. 809. LNCS. Springer, Berlin, Heidelberg, pp. 18–32. DOI: 10.1007/3-540-58108-1_2 |
| Simon | Chapter 2, page 31 | Ray Beaulieu et al. (2013). *The SIMON and SPECK Families of Lightweight Block Ciphers*. Cryptology ePrint Archive, Report 2013/404. URL: https://eprint.iacr.org/2013/404 |
| Rijndael | Chapter 3, page 39 | Joan Daemen and Vincent Rijmen (2020). *The Design of Rijndael – The Advanced Encryption Standard (AES)*. 2nd ed. Information Security and Cryptography. Springer, Berlin, Heidelberg. ISBN: 978-3-662-60768-8. DOI: 10.1007/978-3-662-60769-5. |
| Midori | Chapter 3, page 39 | Subhadeep Banik et al. (Nov. 2015). "Midori: A Block Cipher for Low Energy." In: *ASIACRYPT 2015, Part II*. Ed. by Tetsu Iwata and Jung Hee Cheon. Vol. 9453. LNCS. Springer, Berlin, Heidelberg, pp. 411–436. DOI: 10.1007/978-3-662-48800-3_17 |
| Speck | Chapter 3, page 44 | Ray Beaulieu et al. (2013). *The SIMON and SPECK Families of Lightweight Block Ciphers*. Cryptology ePrint Archive, Report 2013/404. URL: https://eprint.iacr.org/2013/404 |

# References

Ashur, Tomer, Tim Beyne, and Vincent Rijmen (Apr. 2020). "Revisiting the Wrong-Key-Randomization Hypothesis." In: *Journal of Cryptology* 33.2, pp. 567–594. DOI: 10.1007/s00145-020-09343-2.

Baignères, Thomas, Pascal Junod, and Serge Vaudenay (Dec. 2004). "How Far Can We Go Beyond Linear Cryptanalysis?" In: *ASIACRYPT 2004*. Ed. by Pil Joong Lee. Vol. 3329. LNCS. Springer, Berlin, Heidelberg, pp. 432–450. DOI: 10.1007/978-3-540-30539-2_31.

Baignères, Thomas, Jacques Stern, and Serge Vaudenay (Aug. 2007). "Linear Cryptanalysis of Non Binary Ciphers." In: *SAC 2007*. Ed. by Carlisle M. Adams, Ali Miri, and Michael J. Wiener. Vol. 4876. LNCS. Springer, Berlin, Heidelberg, pp. 184–211. DOI: 10.1007/978-3-540-77360-3_13.

Banik, Subhadeep et al. (Nov. 2015). "Midori: A Block Cipher for Low Energy." In: *ASIACRYPT 2015, Part II*. Ed. by Tetsu Iwata and Jung Hee Cheon. Vol. 9453. LNCS. Springer, Berlin, Heidelberg, pp. 411–436. DOI: 10.1007/978-3-662-48800-3_17.

Beaulieu, Ray et al. (2013). *The SIMON and SPECK Families of Lightweight Block Ciphers*. Cryptology ePrint Archive, Report 2013/404. URL: https://eprint.iacr.org/2013/404.

Beierle, Christof, Anne Canteaut, and Gregor Leander (2018). "Nonlinear Approximations in Cryptanalysis Revisited." In: *IACR Transactions on Symmetric Cryptology* 2018.4, pp. 80–101. ISSN: 2519-173X. DOI: 10.13154/tosc.v2018.i4.80-101.

Beyne, Tim (Dec. 2018). "Block Cipher Invariants as Eigenvectors of Correlation Matrices." In: *ASIACRYPT 2018, Part I*. Ed. by Thomas Peyrin and Steven Galbraith. Vol. 11272. LNCS. Springer, Cham, pp. 3–31. DOI: 10.1007/978-3-030-03326-2_1.

— (Dec. 2021). "A Geometric Approach to Linear Cryptanalysis." In: *ASIACRYPT 2021, Part I*. Ed. by Mehdi Tibouchi and Huaxiong Wang. Vol. 13090. LNCS. Springer, Cham, pp. 36–66. DOI: 10.1007/978-3-030-92062-3_2.

— (June 2023). "A Geometric Approach to Symmetric-Key Cryptanalysis." PhD thesis. KU Leuven.

Biryukov, Alex, Christophe De Cannière, and Michaël Quisquater (2004). "On Multiple Linear Approximations." In: *Advances in Cryptology – CRYPTO 2004, 24th Annual International Cryptology Conference, Santa Barbara, California, USA,*

*August 15–19, 2004, Proceedings.* Ed. by Matthew K. Franklin. Vol. 3152. LNCS. Springer, pp. 1–22. DOI: 10.1007/978-3-540-28628-8\_1.

Blondeau, Céline and Kaisa Nyberg (2017). "Joint Data and Key Distribution of Simple, Multiple, and Multidimensional Linear Cryptanalysis Test Statistic and Its Impact to Data Complexity." In: *Designs, Codes and Cryptography* 82, pp. 319–349.

Bogdanov, Andrey et al. (Dec. 2012). "Integral and Multidimensional Linear Distinguishers with Correlation Zero." In: *ASIACRYPT 2012*. Ed. by Xiaoyun Wang and Kazue Sako. Vol. 7658. LNCS. Springer, Berlin, Heidelberg, pp. 244–261. DOI: 10.1007/978-3-642-34961-4_16.

Bogdanov, Andrey and Vincent Rijmen (2014). "Linear Hulls with Correlation Zero and Linear Cryptanalysis of Block Ciphers." In: *DCC* 70.3, pp. 369–383. DOI: 10.1007/s10623-012-9697-z.

Bogdanov, Andrey and Elmar Tischhauser (Mar. 2014). "On the Wrong Key Randomisation and Key Equivalence Hypotheses in Matsui's Algorithm 2." In: *FSE 2013*. Ed. by Shiho Moriai. Vol. 8424. LNCS. Springer, Berlin, Heidelberg, pp. 19–38. DOI: 10.1007/978-3-662-43933-3_2.

Bogdanov, Andrey and Meiqin Wang (Mar. 2012). "Zero Correlation Linear Cryptanalysis with Reduced Data Complexity." In: *FSE 2012*. Ed. by Anne Canteaut. Vol. 7549. LNCS. Springer, Berlin, Heidelberg, pp. 29–48. DOI: 10.1007/978-3-642-34047-5_3.

Collard, Baudoin and François-Xavier Standaert (Apr. 2009). "A Statistical Saturation Attack against the Block Cipher PRESENT." In: *CT-RSA 2009*. Ed. by Marc Fischlin. Vol. 5473. LNCS. Springer, Berlin, Heidelberg, pp. 195–210. DOI: 10.1007/978-3-642-00862-7_13.

Collard, Baudoin, Francois-Xavier Standaert, and Jean-Jacques Quisquater (2007). "Improving the Time Complexity of Matsui's Linear Cryptanalysis." In: *Information Security and Cryptology – ICISC 2007: 10th International Conference, Seoul, Korea, November 29–30, 2007. Proceedings 10*. Springer, Berlin, Heidelberg, pp. 77–88. DOI: 10.1007/978-3-540-76788-6_7.

Daemen, Joan (Mar. 1995). "Cipher and Hash Function Design Strategies Based on Linear and Differential Cryptanalysis." PhD thesis. KU Leuven.

Daemen, Joan, René Govaerts, and Joos Vandewalle (Dec. 1994). "A New Approach to Block Cipher Design." In: *FSE'93*. Ed. by Ross J. Anderson. Vol. 809. LNCS. Springer, Berlin, Heidelberg, pp. 18–32. DOI: 10.1007/3-540-58108-1_2.

— (Dec. 1995). "Correlation Matrices." In: *FSE'94*. Ed. by Bart Preneel. Vol. 1008. LNCS. Springer, Berlin, Heidelberg, pp. 275–285. DOI: 10.1007/3-540-60590-8_21.

Daemen, Joan, Lars R. Knudsen, and Vincent Rijmen (Jan. 1997). "The Block Cipher Square." In: *FSE'97*. Ed. by Eli Biham. Vol. 1267. LNCS. Springer, Berlin, Heidelberg, pp. 149–165. DOI: 10.1007/BFb0052343.

Daemen, Joan and Vincent Rijmen (Dec. 2001). "The Wide Trail Design Strategy." In: *8th IMA International Conference on Cryptography and Coding*. Ed. by Bahram Honary. Vol. 2260. LNCS. Springer, Berlin, Heidelberg, pp. 222–238. DOI: 10.1007/3-540-45325-3_20.

(2020). *The Design of Rijndael – The Advanced Encryption Standard (AES)*. 2nd ed. Information Security and Cryptography. Springer, Berlin, Heidelberg. ISBN: 978-3-662-60768-8. DOI: 10.1007/978-3-662-60769-5.

Halmos, Paul R. (1958). *Finite-dimensional Vector Spaces*. 1st ed. Undergraduate Texts in Mathematics. Springer New York, NY.

Harpes, Carlo, Gerhard G. Kramer, and James L. Massey (May 1995). "A Generalization of Linear Cryptanalysis and the Applicability of Matsui's Piling-Up Lemma." In: *EUROCRYPT'95*. Ed. by Louis C. Guillou and Jean-Jacques Quisquater. Vol. 921. LNCS. Springer, Berlin, Heidelberg, pp. 24–38. DOI: 10.1007/3-540-49264-X_3.

Harpes, Carlo and James L. Massey (Jan. 1997). "Partitioning Cryptanalysis." In: *FSE'97*. Ed. by Eli Biham. Vol. 1267. LNCS. Springer, Berlin, Heidelberg, pp. 13–27. DOI: 10.1007/BFb0052331.

Hermelin, Miia, Joo Yeon Cho, and Kaisa Nyberg (2008). "Multidimensional Linear Cryptanalysis of Reduced Round Serpent." In: *Information Security and Privacy, 13th Australasian Conference, ACISP 2008, Wollongong, Australia, July 7–9, 2008, Proceedings*. Ed. by Yi Mu, Willy Susilo, and Jennifer Seberry. Vol. 5107. LNCS. Springer, pp. 203–215. DOI: 10.1007/978-3-540-70500-0\_15.

Kaliski Jr., Burton S. and Matthew J. B. Robshaw (Aug. 1994). "Linear Cryptanalysis Using Multiple Approximations." In: *CRYPTO'94*. Ed. by Yvo Desmedt. Vol. 839. LNCS. Springer, Berlin, Heidelberg, pp. 26–39. DOI: 10.1007/3-540-48658-5_4.

Knudsen, Lars R. and Matthew J. B. Robshaw (May 1996). "Non-Linear Approximations in Linear Cryptanalysis." In: *EUROCRYPT'96*. Ed. by Ueli M. Maurer. Vol. 1070. LNCS. Springer, Berlin, Heidelberg, pp. 224–236. DOI: 10.1007/3-540-68339-9_20.

Kullback, Solomon and Richard A. Leibler (1951). "On Information and Sufficiency." In: *The Annals of Mathematical Statistics* 22.1, pp. 79–86.

Leander, Gregor et al. (Aug. 2011). "A Cryptanalysis of PRINTcipher: The Invariant Subspace Attack." In: *CRYPTO 2011*. Ed. by Phillip Rogaway. Vol. 6841. LNCS. Springer, Berlin, Heidelberg, pp. 206–221. DOI: 10.1007/978-3-642-22792-9_12.

Leander, Gregor, Brice Minaud, and Sondre Rønjom (Apr. 2015). "A Generic Approach to Invariant Subspace Attacks: Cryptanalysis of Robin, iSCREAM and Zorro." In: *EUROCRYPT 2015, Part I*. Ed. by Elisabeth Oswald and Marc Fischlin. Vol. 9056. LNCS. Springer, Berlin, Heidelberg, pp. 254–283. DOI: 10.1007/978-3-662-46800-5_11.

Matsui, Mitsuru (May 1994a). "Linear Cryptanalysis Method for DES Cipher." In: *EUROCRYPT'93*. Ed. by Tor Helleseth. Vol. 765. LNCS. Springer, Berlin, Heidelberg, pp. 386–397. DOI: 10.1007/3-540-48285-7_33.

(Aug. 1994b). "The First Experimental Cryptanalysis of the Data Encryption Standard." In: *CRYPTO'94*. Ed. by Yvo Desmedt. Vol. 839. LNCS. Springer, Berlin, Heidelberg, pp. 1–11. DOI: 10.1007/3-540-48658-5_1.

Nyberg, Kaisa (May 1995). "Linear Approximation of Block Ciphers (Rump Session)." In: *EUROCRYPT'94*. Ed. by Alfredo De Santis. Vol. 950. LNCS. Springer, Berlin, Heidelberg, pp. 439–444. DOI: 10.1007/BFb0053460.

Schulte-Geers, Ernst (2013). "On CCZ-equivalence of Addition mod $2^n$." In: *Designs, Codes and Cryptography* 66, pp. 111–127.

Selçuk, Ali Aydin (Jan. 2008). "On Probability of Success in Linear and Differential Cryptanalysis." In: *Journal of Cryptology* 21.1, pp. 131–147. DOI: 10.1007/s00145-007-9013-7.

Tardy-Corfdir, Anne and Henri Gilbert (Aug. 1992). "A Known Plaintext Attack of FEAL-4 and FEAL-6." In: *CRYPTO'91*. Ed. by Joan Feigenbaum. Vol. 576. LNCS. Springer, Berlin, Heidelberg, pp. 172–181. DOI: 10.1007/3-540-46766-1_12.

Terras, Audrey (1999). *Fourier Analysis on Finite Groups and Applications*. London Mathematical Society Student Texts. Cambridge University Press, Cambridge.

Todo, Yosuke, Gregor Leander, and Yu Sasaki (Dec. 2016). "Nonlinear Invariant Attack – Practical Attack on Full SCREAM, iSCREAM, and Midori64." In: *ASIACRYPT 2016, Part II*. Ed. by Jung Hee Cheon and Tsuyoshi Takagi. Vol. 10032. LNCS. Springer, Berlin, Heidelberg, pp. 3–33. DOI: 10.1007/978-3-662-53890-6_1.

Vaudenay, Serge (1996a). "An Experiment on DES Statistical Cryptanalysis." In: *CCS '96, Proceedings of the 3rd ACM Conference on Computer and Communications Security, New Delhi, India, March 14–16, 1996*. Ed. by Li Gong and Jacques Stearn. ACM, New York, pp. 139–147. DOI: 10.1145/238168.238206.

— (Mar. 1996b). "An Experiment on DES Statistical Cryptanalysis." In: *ACM CCS 96*. Ed. by Li Gong and Jacques Stern. ACM Press, New York, pp. 139–147. DOI: 10.1145/238168.238206.

Wallén, Johan (Feb. 2003). "Linear Approximations of Addition Modulo $2^n$." In: *FSE 2003*. Ed. by Thomas Johansson. Vol. 2887. LNCS. Springer, Berlin, Heidelberg, pp. 261–273. DOI: 10.1007/978-3-540-39887-5_20.

# Index

3-Way, 3

absolutely continuous, 92
acceptance region, 93
active S-box, 25
add-rotate-xor (ARX), 44
adjoint, 138
Advanced Encryption Standard (AES)
   block size of, 4
   standardization of, 47
advantage
   of hypothesis test, 54
   of key-recovery attack, 62
affine
   function, 22
   subspace, 122
algebraic normal form, 125
almost everywhere, 92
alternative hypothesis, 54
analysis phase, 65
angles
   between vector spaces, 148
   between vectors, 137
annihilator
   subgroup, 145
   subspace, 148, 159
anti-isomorphism, 136
approximation map
   *definition of*, 161
   geometry of, 161–162
   projection functions, 166
attack, difficulty of, 2

balanced
   Boolean function, 29

projection function, 128, 165
Bayes factor, 99–100
best approximation theorem, 138
bias
   empirical bias, 10
   linear approximation, 4–5
   random bit, 8, 17
bilinear function, 81, 139
binomial distribution, 61, 169
bit permutation, 3, 7
bitvector, 1
block cipher, 1
   block size, 1, 4
   design of, 2–4
   example cipher, 3–4
   Feistel cipher, 31, 117
   inner and outer part, 12, 64
   *list of*, 173
   Rijndael-like, 39–41
block code, 49
Boolean function, 18
branch and bound, 34–38
branch number, 40, 49
bricklayer function, 22

capacity, 83
   average, 115
   *definition of*, 74
   from principal correlations, 162
carry, 46
cell, 39
central limit theorem, 58, 61, 75, 169, 170
$\chi^2$ test, 84, 87, 88
chosen-plaintext attack
   *definition of*, 2

178

multidimensional linear, 85, 157
zero-correlation linear, 116
circulant matrix
  convolution, 73
  *definition of*, 69
  diagonalization, 71
  multiplication, 70–72
coalgebra, 166
collision, 164
combinatorial optimization, 34
complement
  algebraic, 148, 161
  orthogonal, 81, 137, 148, 161
composition
  of correlation matrices, 20, 156
  of round functions, 2
  of transition matrices, 153
  using piling-up, 8
concatenation, 3, 22
conjunctive normal form, 44
convex
  hull, 42
  polytope, 42
  set, 42
convolution theorem
  Fourier transformation, 73
  piling-up lemma, 18
correlation
  Boolean functions, 18–19
  coefficient, 18
  empirical correlation, 57
  linear approximation, 19
  principal correlation, 161
  random bit, 17–18
correlation matrix
  affine function, 22
  bitwise-and function, 31
  *definition of*, 19, 155
  geometric approach to, 155–157
  group homomorphism, 156
  modular addition, 46
  quadratic function, 31–32
  random function, 103, 106
  random permutation, 106
cost function, 93, 99
counter mode, 61
covariance
  empirical correlations, 75–76
  matrix, 75, 95, 106, 170
cryptanalytic property, 150–152

cryptographic primitives, *see* primitives
cube function, 164

Data Encryption Standard (DES)
  linear cryptanalysis, 14
  S-box $S_5$, 15
data-complexity
  chosen-plaintext, 85, 116
  *formulary*, 171
  geometric approach, 154, 162
  known correlation, 58, 77, 84
  unknown correlation, 59, 78, 84
  unknown signs, 80, 101–102
  zero-correlation, 115–116
degree
  Boolean function, 125
  polynomial, 49, 164
degrees of freedom, 88
depth-first traversal, 34–36
differential cryptanalysis, 150
direct sum
  groups, 141, 142
  vector spaces, 137, 161
distillation phase, 64
distinguisher, 2, 54
dominant trail, 25, 163
dot product
  bitvectors, 81, 82
  $\mathbb{C}[G]$ or $\mathbb{C}^G$, *see* inner product
dual
  basis, 143
  Pontryagin dual group, 140–142
  vector space, 134–136
dummy variable, 41

effective linear approximation, 4
encrypt-mix-encrypt, 118
errors of the 1st and 2nd type, 93
estimator, 52–54, 154
Euclidean norm, 21, 134, 153
evaluation map, 131, 135, 141
exhaustive search, 2, 65, 72

false-positive probability
  average, 99
  *definition of*, 54
  invariant subspace, 122
  trade-off, 93
  zero-correlation linear, 111, 114
fast Fourier (FFT) method, 68–72

Feistel cipher
  definition of, 31, 117
  zero-correlation, 117
finite field
  exponential sum over $\mathbb{F}_q$, 164
  inversion in $\mathbb{F}_{2^b}$, 50
  linear code over $\mathbb{F}_{2^b}$, 49
  of order two, $\mathbb{F}_2$, 1
  trace function, 50
fixed point, 164
forking operation, 15
formal linear combination, 133
Fourier transformation, 18, 83, 143–144
free vector space, 133
Frobenius norm, 166
fundamental theorem of finite Abelian groups, 142

Galois Counter Mode, 61
geometric approach, 150–154
graph, 34
graph of a function, 45
greedy search, 35
group
  action, 140
  annihilator, 145
  characters, 140–143
  linear cryptanalysis on, 155
  Pontryagin dual, 140

Hamming
  distance between functions, 15
  weight of a bitvector, 40, 47
  weight of codewords, 49
hypergeometric distribution
  multivariate, 114
  univariate, 61
hypothesis testing
  composite, 98–104
  definition of, 54–56
  Fisher, 54
  multivariate normal, 94–95
  nearly equal, 95–98
  Neyman–Pearson, 54, 93–94
  simple, 92–98

I/O sum, *see also* nonlinear approximation, 129
information of discrimination, 96
inner product
  between functions, 19, 28, 136

induced norm, 136
inner product space, 136–138
integral cryptanalysis, 122, 150
integration
  of measurable functions, 91
  by parts, 79, 169
invariant
  black-box approach, 124
  forward and backward, 160
  geometric approach to, 159–160
  nonlinear invariant, 124–127, 159
  subspace, 122–124
  subspace (linear algebra), 149
  symmetry, 122
inversion in a finite field, 50
isometry, 135
iterated cipher, 2

Jacobian determinant, 170
Jeffrys divergence, 96

key, 1
  addition of, 3, 7, 24
  candidate key, 12
  difference invariant bias, 119
  expanded key, 2
  key-alternating cipher, 2–3
  ranking, 61
  recovery, 1, 9, 64–72, 86, 111
  round key, 3
  schedule, 2, 45, 68
  weak key, 123
Kloosterman sum, 50
known-plaintext attack
  data-complexity, *see also* data-complexity, 162
  definition of, 2, 11
Kronecker product, 23, 140
Kullback–Leibler divergence, 96

Lai–Massey construction, 90
lexicographic ordering, 21
likelihood ratio
  average and variance of, 96
  definition of, 94
  logarithmic, 94
  multivariate normal, 94–95
linear algebra, 133–140
linear approximation
  definition of, 4

geometric approach to, 155–156
multidimensional, 81–83
multiple, 74–76
table (LAT), 5
zero-correlation, 107
linear branch number, *see* branch number
linear discriminant analysis, 95
linear function, 14, 22
linear functional, 135
linear hull theorem, 33
linear programming, 38
linear trail, *see* trail
logarithmic likelihood ratio, 94
lookup table, 3

mask, 4
Matsui's algorithm
    algorithm 1, 10–11
    algorithm 2, 12–13, 64–68
    trail search, 36–38
maximum distance separable (MDS)
    *construction of*, 49
    *definition of*, 49
measure space, 91
metric space, 133
miss-in-the-middle, 108, 110, 159
MixColumns, 40
mixed-integer linear programming
    CPLEX, 43
    *definition of*, 39
    Google OR-Tools, 50
    Gurobi, 43
    LP format, 43
model error, 102, 171
modeling phase, 64
modular addition, 44
most powerful on average, 99
multidimensional linear
    approximation, 81–83, 127, 128
    cryptanalysis, 81–85, 157
multiple linear
    approximation, 74–76, 157
    cryptanalysis, 74–81
    key-recovery, *impact on*, 86
multivariate
    central limit theorem, 75, 106, 170
    hypergeometric distribution, 114
    hypothesis testing, 74, 94–95
    normal distribution, 75, 94–95, 169–170

Neyman–Pearson lemma, 93
nonlinear
    approximation, 127–128, 159
    trail, 127
nonlinearity, 15
norm
    *definition of*, 134
    dual norm, 135
    Euclidean, 21, 153
    Euclidean norm, 134
    normed vector space, 134–136
    $p$-norm, 134
normal distribution
    background on, 168–170
    density function $\varphi$, 168
    distribution function $\Phi$, 55, 168
    fourth moment, 79, 88
    hypothesis test on mean, 55–56
    mixture, 100
    multivariate, 75, 94–95, 169–170
null hypothesis, 54

objective function, 38
order statistic, 62
orthogonal
    complement, 81, 137, 148, 161
    matrix, 20, 125
    projection, 137
orthogonality
    correlation matrices, 21
    group characters, 142
    vectors, 137
orthonormal basis, 137

$p$-norm, 134
partitioning cryptanalysis, 128
piling-up lemma, 8, 18
pointwise product, 140
Poisson summation formula, 82
Pontryagin duality, 140, 145–146, 149
posterior distribution, 100, 103
power, 93
primitives
    *analysis of*, 4
    block ciphers, *see* block cipher
primitives, *analysis of*, 1
principal angle, 149, 162
principal correlation, 161
prior distribution, 99, 100, 171
probability measure, 92

# 182  Index

projection
  framework, 128
  function, 128, 158, 165
  linear, 83, 95
  onto quotient, 81, 145
  orthogonal, 137
pruning, 68, 70
pullback operator, 152
pushforward operator, 152
Pythagorean theorem, 137

quadratic
  Boolean function, 32, 125
  form, 31
  nonlinear invariant, 125
quadratic discriminant analysis, 95
quickhull algorithm, 42
quotient
  of groups, 145
  of vector spaces, 81

Radon–Nykodym density, 92
Reed–Solomon code, 49
right key
  *definition of*, 12
  randomization, 99–102
round function, 2

S-box
  active S-box, 25
  *definition of*, 3
  DES, 15
  layer, 3, 22
  Rijndael, 50
sample
  average, 53
  with replacement, 53, 61
  without replacement, 61, 104
  *use of*, 52
satisfiability (SAT), 44
satisfiability modulo theories (SMT)
  Boolector, 47
  *definition of*, 44
  LibSMT, 47
  PySMT, 50
  Z3, 47
saturation
  attack, 120–122
  property, 114, 121, 158
  statistical attack, 85, 122, 127
  zero-correlation, 159

search phase, 65
security, *definition of*, 1
self-adjoint, 138, 149
self-dual, 136
separating hyperplane, 95
set cover problem, 42
ShiftRows, 40
shortest path, 35
Simon, 31
simple hypothesis, 92
simple model, 57
simplex algorithm, 39
singleton bound, 49
singular value, 138–139, 149, 162, 166
SPC, 90
Speck, 44
squared Euclidean imbalance, 83, 98, 127, 166
standard basis, 133, 134
statistical
  attack, 10
  inference, 52–56
  saturation attack, 85, 122
Stirling's approximation, 113
SubCells, 40
substitution-permutation network, 3
success probability, 2
  *definition of*, 54
  *formulary*, 171
  invariant subspace, 122
  known correlation, 58, 77
  trade-off, 93
  unknown correlation, 59, 78
  unknown signs, 80
  zero-correlation, 107, 115–116

Taylor series, 97
tensor
  elementary, 139
  rank-one, 139
tensor product
  of linear maps, 140
  of vector spaces, 139–140
Thue–Morse sequence, 40
trace
  finite fields, 50
  of a matrix, 164
trail
  geometric approach, 162
  linear trail, 7–9, 24–26
transition matrix, 152
transpose of a linear map, 144

tree, 34
true-positive probability, 54

uniformly most powerful, 93, 99
universal property, 139

variance, 53

Walsh–Hadamard transformation, 28
weak key, 123
Weil's bound, 164
wide-trail design strategy, 39

wrong key
   *definition of*, 12
   randomization, 57, 67, 99, 102–104

zero-correlation
   geometric approach, 159
   linear approximation, 107
   linear cryptanalysis, 107–116
   miss-in-the-middle, 108, 159
   multidimensional, 113–115
   multiple, 115
   property, 159
   random permutation, 113
   statistical approach, 115

For EU product safety concerns, contact us at Calle de José Abascal, 56–1°, 28003 Madrid, Spain or eugpsr@cambridge.org.